MODEL ANSWERS

IB BIOLOGY

SECOND EDITION

This model answer booklet is a companion publication to provide answers for the activities in the IB Biology Student Workbook (2nd edition). These answers have been produced as a separate publication to keep the cost of the workbook itself to a minimum. All answers to set questions are provided, but chapter reviews are the student's own and no model answer is set. Working and explanatory notes have been provided where clarification of the answer is appropriate.

www.the**BIOZONE**.com

ISBN 978-1-927173-94-7

Published by **BIOZONE** International Ltd
Sixth printing with corrections

BIOZONE

USA, CANADA & REST OF WORLD:
BIOZONE International Ltd.
PO Box 5002, Hamilton 3242, New Zealand
Telephone: +64 7-856 8104
Fax: +64 7-856 9243
Toll FREE phone: 1-855-246-4555 (USA-Canada only)
Toll FREE fax: 1-855-935-3555 (USA-Canada only)
Email: sales@biozone.co.nz
Website: www.the**BIOZONE**.com

Contents

Cell Biology

Molecular Biology

Genetics

Contents

Contents

1. Cell Theory (page 5)

1. Cells are the smallest units of any living thing that display the all the functions of life. Below this level no particular part shows all the functions of life.
2. Organisms such a *Caulerpa* show multicellular characteristics while being one large cell. Cells do not divide into new cells in the conventional sense.

2. Unicellular Organisms (page 6)

1. *Scenedesmus* defends itself against grazers by (1) living in a colony and (2) the presence of protective spines.
2. These numbers are exponential increases of 2. Cells divide producing a larger colony. Above 16 cells the colony is unstable and divides in two.
3. (a) *Paramecium* is a predator. It gains nutrients (food) by actively hunting other single celled organisms.
   (b) *Scenedesmus* is photosynthetic. It gains food by using sunlight to convert carbon dioxide into glucose.
4. *Paramecium* needs to be mobile in order to find food.

3. Surface Area and Volume (page 7)

1.

Cube	Surface area	Volume	Ratio
3 cm:	3 x 3 x 6 = 54	3 x 3 x 3 = 27	2.0 to 1
4 cm:	4 x 4 x 6 = 96	4 x 4 x 4 = 64	1.5 to 1
5 cm:	5 x 5 x 6 = 150	5 x 5 x 5 = 125	1.2 to 1

2. Surface area to volume graph:

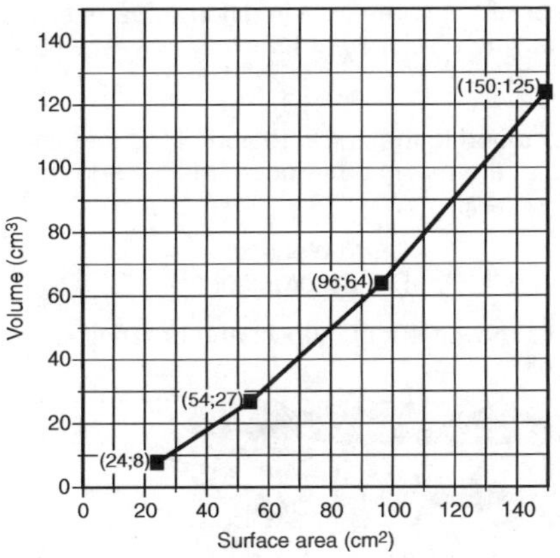

3. Volume
4. Increasing size leads to less surface area for a given volume. The surface area to volume ratio decreases.
5. (a)

Cube	1	2	3
Total volume	1	8	64
Volume not pink	0.125 (0.5^3)	3.375 (1.5^3)	42.8 (3.5^3)
Diffused volume	0.875	4.625	21.2
Percentage diffusion	87.5	57.8	33.1

   (b) An increase in cell size reduces the ability of diffusion to transport molecules into and out of a cell at the rate required to maintain the cell's original metabolic rate. (i.e. volume increases faster that surface area and thus the proportion of molecules diffusing across the plasma membrane decreases in relation to the size of the cell).
6. A cell measuring 1 cm x 1 cm x 1 cm has a surface area of 6 cm^3. Eight of them will have a volume of 8 cm^3 but a surface area of 42 cm^2. A cell measuring 2 cm x 2 cm x 2 cm will have a volume of 8 cm^3 but a surface area of only 24 cm^2. Thus eight small cells are more able to acquire nutrients due to their larger surface area to volume ratios.

4. Cell Sizes (page 9)

1. (a) *Daphnia*: 1800 µm 1.8 mm
   (b) *Giardia:* 15 µm 0.015 mm
   (c) Nucleus 10 µm 0.01 mm
   (d) *Elodea:* 47 µm 0.047 mm
   (e) Chloroplast 7 µm 0.007 mm
   (f) *Paramecium*: 250 µm 0.25 mm
2. (a) Chloroplast < nucleus < *Giardia* < *Elodea* < *Paramecium* < *Daphnia*
   (b) Visible to naked eye: *Daphnia*
3. (a) 0.00025 mm
   (b) 0.45 mm
   (c) 0.0002 mm

5. Calculating Linear Magnification (page 10)

1. Actual size = image size ÷ magnification
   = 52,000 µm ÷ 140
   = 371 µm (0.37 mm)
2. (a) Actual length of scale line = 10 mm
   Given length of scale line = 0.5 mm
   10 ÷ 0.5 = 20 x magnification
   (b) Measured length = 42 mm
   Magnification = 20 x
   Actual length = 42 ÷ 20 = 2.1 mm
3. 43 mm = 43,000 µm
   Magnification = size of the image ÷ actual size of object
   = 43,000 µm ÷ 2 µm
   = 21,500 x magnification

6. Multicellularity (page 11)

1. Emergent properties are properties displayed by the whole that are not evident in the components. For example, the properties of metabolism and life emerge from all the combined individual biochemical reactions in cells. Muscle contraction emerges from the combined activity of individual muscle cells.
2. Different cells, with different functions produce different kinds of tissue specialized to perform different tasks, e.g. muscle tissue contracts, blood transports materials around the body.

7. Stem Cells and Differentiation (page 12)

1. (a) Self-renewal: they are able to divide multiple times while maintaining an undifferentiated state.
   (b) Potency: they can differentiate into specialized cell types.
2. Stem cells differentiate to give rise to all the cell types of the multicellular organism. The zygote is totipotent and can differentiate into any cell type, including extra-embryonic cells. As development proceeds, the cells become pluripotent. They can give rise to most cell types, but not extra-embryonic cells, and are important in the early development of tissues and organs. Throughout life, multipotent adult stem cells maintain these tissues and organs and are committed to produce the different cell types related to the tissue of origin.

8. Types of Stem Cells (page 13)

1. (a) Embryonic stem cells are pluripotent and can form over 220 cell types in the three primary germ layers (ectoderm, endoderm, mesoderm). Adult stem cells are multipotent and can divide only into a limited number of cell types, mainly those of the blood, heart and nerves.
   (b) Embryonic stem cells have vast potential for medical application as they can theoretically be used to replace most damaged cell and tissue types. The medical use of adult stem cells is more limited than ESC, but because there are fewer ethical issues involved ASC are already being used to treat medical problems such as leukemia.

ISBN: 978-1-927173-94-7

2. (a) Answer is based on the student's own reasoned opinion.
   (b) Therapeutic cloning allows new organs to be grown that will not be rejected by the patient's immune system. Adult stem cells can only be used in specific circumstances and can only be used to repair tissues to that type of cell.
3. If a disease is genetic, treating it with ESC from the person's cord blood will not help as the DNA is already damaged.

---

9. Using Stem Cells to Treat Disease (page 15)

1. Embryonic stem cells can be grown into new tissues that can be used to replaced damaged ones.
2. (a) Stem cells can be cultured to develop into retinal pigment epithelium cells, which can be injected into the eye to replace damaged cells.
   (b) The patient's cells will carry the defect which will reappear in the new cells, so the treatment will not be effective.
   (c) Using a patient's own cells means that there is no chance of immune rejection of the tissue. This is useful when replacing an organ damaged by a non-genetic disease.

---

10. Comparing Prokaryotic and Eukaryotic Cells (page 16)

1. An absence of a membrane bound nucleus or organelles. Cell wall made of peptidoglycan. 70S ribosomes.
2. The presence of a membrane-bound nucleus and organelles. Contain 80S ribosomes.
3. Animal cells, plant cells, fungal cells, protists.
4. A: Prokaryote. No visible organelles, e.g a nucleus.
   B: Eukaryote. Clearly visible organelles, including a nucleus.
   C: Eukaryote. Nucleus clearly visible in the centre of the cell.
   D: Prokaryote. No organelles are visible.

---

12. Prokaryotic Cell Structure (page 18)

1. The bacterial cell wall is a semi-rigid structure outside the plasma membrane. It composed of the macromolecule peptidoglycan, with varying amounts of lipopolysaccharides and lipoproteins.
2. Flagella enable bacterial movement out of unsuitable conditions to preferred conditions. Fimbriae are shorter, straighter, and thinner than flagella. They are used for attachment rather than locomotion.
3. Sex pili join two bacteria and allow exchange of genetic material between them.
4.

Structure of a generalized prokaryotic cell

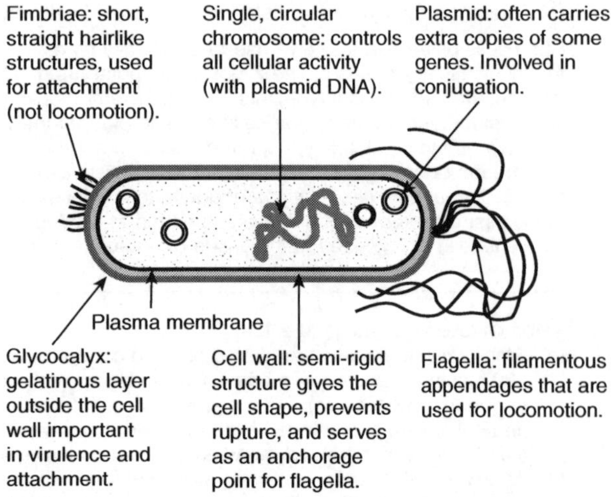

---

13. Binary Fission in Prokaryotes (page 19)

1. Binary fission is a form of asexual reproduction resulting in the division of a cell into two identical cells.
2. The cross wall is an in-growth of the cell wall and cell membrane of a cell dividing by binary fission. When its growth is completed, the cross wall completely divides the cell, resulting in the formation of two cells.
3. The generation time is the time taken for a population of bacterial cells to double.
4. Completed table below:

Min	No.	Min	No.	Min	No.
0	1	140	128	260	8,192
20	2	160	256	280	16,384
40	4	180	512	300	32,768
60	8	200	1,024	320	65,536
80	16	220	2,048	340	131,072
100	32	240	4,096	360	262,144
120	64				

5. (a) 8 (b) 512 (c) 262,144

---

14. Plant Cells (page 20)

1. (a) The plant cell wall provides support for the cell and (through cell turgor) the plant as a whole. It also limits the volume of the plant cell.
   (b) The cell wall provides support for the cell. It is fully permeable and composed of cellulose. The plasma membrane is selectively permeable and composed of a lipid bilayer in which proteins are embedded. It regulates the entry and exit of materials into and out of the cell.
2. Organelles allow parts of the cell to be structurally and functionally specialized for certain roles. They enable multiple processes to occur at the same time in different parts of the cell, thereby increasing efficiency of function.
3. Chloroplast
4. Palisade cells are found in the upper region of the leaf so they are exposed to the maximum amount of light and not shadowed by other cells.
5. A: Nucleus C: Nucleus
   B: Cell wall D: Chloroplasts
6. (a) The directional movement of cytosol and organelles around the cell.
   (b)

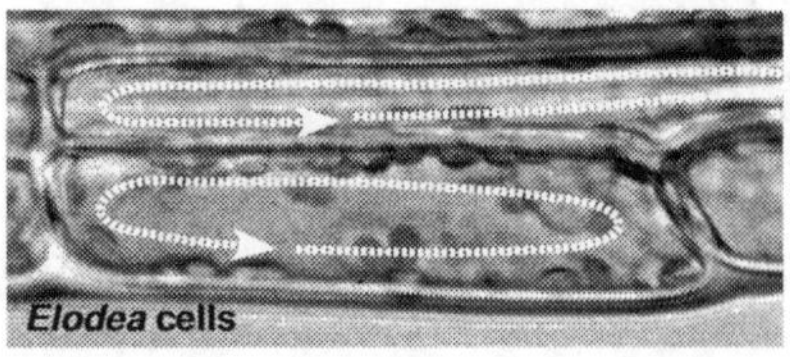

Elodea cells

7. Any three of the following:
   - Starch (branched carbohydrate) granules stored in amyloplasts (energy store).
   - Chloroplasts, discrete plastids containing the pigment chlorophyll, involved in photosynthesis.
   - Large vacuole, often central (vacuoles are present in animal cells, but are only small).
   - Cell wall of cellulose forming the rigid, supporting structure outside the plasma membrane.
   - Plasmodesmata.

---

15. Animal Cells (page 22)

1. A generalized cell exhibits most of the characteristic features of a relatively unspecialized cell. It is not intended to represent any particular cell type but to illustrate features that collectively can be found in a range of cells.
2. (a) The liver contains the organelles that are common to unspecialized cells, e.g. vacuoles, mitochondria, Golgi, smooth and rough ER, peroxisomes, and lysosomes.
   (b) Liver cells contain many mitochondria and large amounts of rough endoplasmic reticulum.
3. A: Nucleus B: Plasma membrane C: Nucleus

ISBN: 978-1-927173-94-7

4. (a)

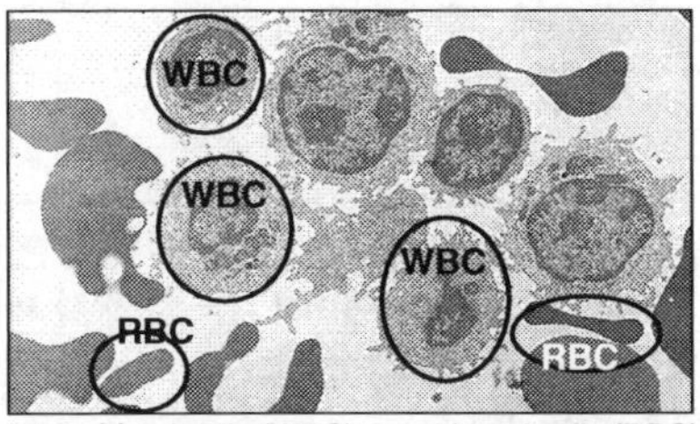

White blood cells (WBC) & red blood cells (RBC)

(b) Any of: The RBCs have no nucleus and they are smaller than the white blood cells. The white blood cells have extensions of the plasma membrane (associated with being mobile and phagocytic), are larger than the RBCs, and have a nucleus.

5. – Centrioles (although these are present in lower plants, they are absent from higher plants). They are microtubular structures responsible for forming the poles and the spindles during cell division.
– Desmosomes. These are points of contact between the plasma membranes of neighbouring cells, which allow cells to combine together to form tissues.

16. Identifying Structures in an Animal Cell (page 24)

1. (a) Plasma membrane
(b) Golgi apparatus
(c) Centriole (TS)
(d) Mitochondrion
(e) Lysosome
(f) Nucleus
(g) Rough ER
(h) Cytoplasm

2. Centrioles

3. Plant cells are enclosed by a rigid cellulose cell wall and do not have the capacity for motility or phagocytosis in the way that animal cells do.

4. (a) Protein production and secretion indicated by a large amount of ER and an extensive Golgi.
(b) Large number of mitochondria indicate that it is metabolically very active (high respiration rate).

5. It has a membrane-bound nucleus and organelles.

6.

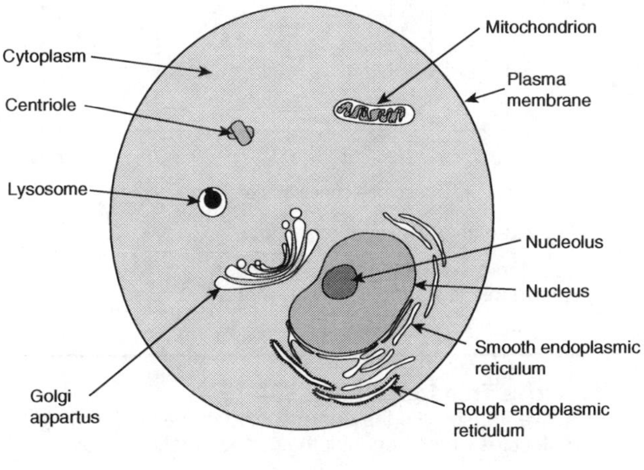

17. Identifying Structures in a Plant Cell (page 25)

1. (a) Cytoplasm
(b) Vacuole
(c) Starch granule
(d) Chloroplast
(e) Mitochondria
(f) Cell wall
(g) Chromosome
(h) Nuclear membrane
(i) Endoplasmic reticulum
(j) Plasma membrane

2. 9 cells (1 complete cell, plus the edges of 8 others)

3. Chloroplasts, cell wall, highly geometric cell shape.

4. (a) Cytoplasm located between the plasma membrane and nuclear membrane (extranuclear).
(b) Composition of cytoplasm: A watery soup of dissolved substances. In eukaryotic cells, organelles are found in the cytoplasm. Cytoplasm = cytosol (including cytoskeleton) + organelles.

5. (a) Starch granules, which occur in specialized plastids called leukoplasts. Starch granules are non-living inclusions, deposited as a reserve energy store.
(b) Vacuoles, which are fluid filled cavities bounded by a single membrane. Plant vacuoles contain cell sap, an aqueous solution of dissolved food material, ions, waste products, and pigments. **Note**: Young plant cells usually have several small vacuoles, which unite in a mature cell to form a large, permanent central vacuole.

6.

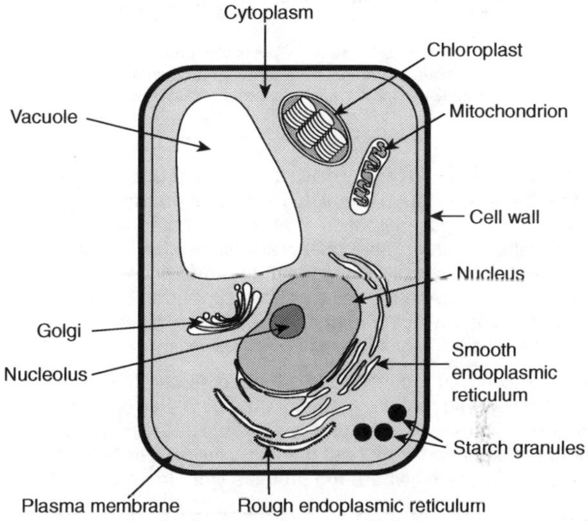

18. The Structure of Membranes (page 26)

1. (a) Channel and carrier proteins
(b) Carrier proteins
(c) Glycoproteins, glycolipids
(d) Cholesterol (and also type of fatty acid present)

2. Phospholipids have hydrophilic heads and hydrophobic tails. The molecules orientate so that the hydrophilic head is towards the exterior of the membrane and the hydrophobic tail is towards the interior.

3. (a) Membranes are composed of a phospholipid bilayer in which are embedded proteins, glycoproteins, and glycolipids. The structure is relatively fluid and the proteins are able to move within this fluid matrix.
(b) This model accounts for the properties we observe in cellular membranes: their fluidity (how components move within the membrane relative to one-another) and their mosaic nature (the way in which the relative proportions of the membrane components, i.e. proteins, glycoproteins, glycolipids etc., can vary from membrane to membrane). The fluid mosaic model also accounts for how membranes can allow for the selective passage of materials (through protein channels for example) and how they enable cell-cell recognition (again, as a result of membrane components such as glycoproteins).

4. The plasma membrane forms the outer limit of the cell, contains proteins that confer cellular recognition and controls the entry and exit of materials in and out of the cell. Intracellular membranes keep the cytoplasm separate from the extracellular spaces and provide compartments for metabolic reactions and a surface for attaching enzymes.

5. (a) Golgi, mitochondria, chloroplasts, nucleus.
(b) Membranes compartmentalize the location of reactions, control the entry and exit of substances, and provide an enzyme attachment surface.

6. Cholesterol acts as a packing molecule, interacting with the phospholipids to regulate the consistency of the membrane (preventing both crystallization and excessive fluidity).

ISBN: 978-1-927173-94-7

7. (a)-(c) in any order: oxygen, food (glucose), minerals and trace elements, water.
8. (a) Carbon dioxide (b) Nitrogenous wastes
9. Plasma membrane:

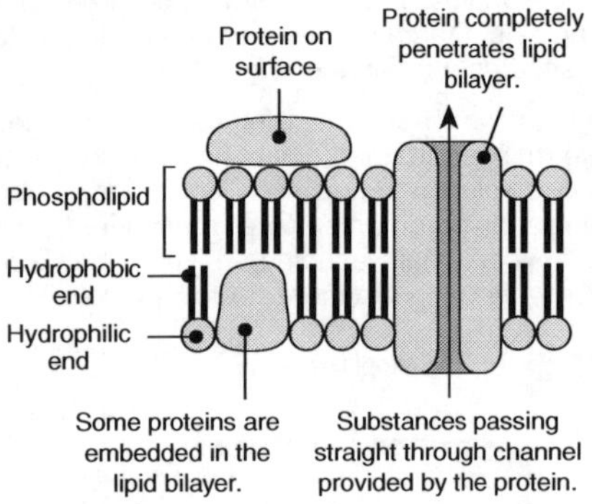

19. How Do We Know? Membrane Structure (page 28)

1. The impressions left in one side of the membrane after freeze fracture give evidence of proteins located within the membrane. Fracturing the membrane allowed scientists to observe the presence of integral membrane proteins which span the membrane lipid bilayer. This supported the fluid mosaic model, in which membrane-bound proteins are able to move relatively freely within the membrane.
2. If the bilayer had a continuous protein coat, the freeze fracture specimen would look flat and uniform when viewed under the electron microscope. The proteins are discrete complexes randomly spaced throughout the lipid bilayer. The bumps observed indicates where the proteins were located.

20. Diffusion (page 29)

1. Diffusion is the passive movement of particles down a concentration gradient (from high to low concentration).
2. They involve movement of substances across a membrane with no input of energy (they are passive).
3. (a) The salt ions would diffuse into the water faster (the salt would dissolve faster).
   (b) The salt ions will diffuse more slowly in the solution.

21. Osmosis (page 30)

1. Osmosis is the diffusion of water molecules across a partially permeable membrane from a region of lower solute concentration (higher free water molecule concentration) to a region of higher solute concentration (lower free water molecule concentration).
2. (a) The net movement of water is from right to left:

   ←

   (b) The water moved into the dialysis tubing because it contained the sucrose solution, and therefore had a higher solute concentration and a lower concentration of free water molecules. The water moved down its concentration gradient.
3. The height of the water would increase.

22. Estimating Osmolarity (page 31)

1. Completed table (totals section only)

0.00 Mol L^{-1}		Initial mass (g)	Final mass (g)
Total		**15.46**	**17.22**
Change (g)	**1.76**		
% Change	**11.38%**		
0.25 Mol L^{-1}		Initial mass (g)	Final mass (g)
Total		**19.18**	**17.93**
Change (g)	**-1.25**		
% Change	**-6.52%**		
0.50 Mol L^{-1}		Initial mass (g)	Final mass (g)
Total		**18.26**	**16.14**
Change (g)	**-2.12**		
% Change	**-11.61%**		
0.75 Mol L^{-1}		Initial mass (g)	Final mass (g)
Total		**19.16**	**13.99**
Change (g)	**-5.17**		
% Change	**-26.98%**		
1.00 Mol L^{-1}		Initial mass (g)	Final mass (g)
Total		**16.14**	**13.00**
Change (g)	**-3.14**		
% Change	**-19.45%**		

2.

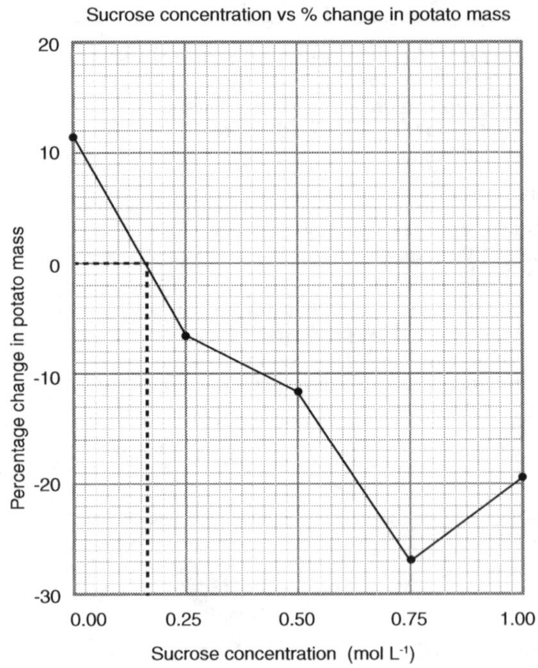

3. Approximately 0.16 mol L^{-1}
4. Hypotonic: 0.00 mol L^{-1}
   Hypertonic: 0.25 mol L^{-1}, 0.50 mol L^{-1}, 0.75 mol L^{-1}
   1.00 mol L^{-1}

23. Active Transport (page 32)

1. Active transport is the energy using process of moving molecules or ions against their concentration gradient.
2. ATP.
3. Primary active transport uses energy gained directly from ATP. Secondary active transport uses energy in the form of a concentration gradient formed by transport proteins using ATP for energy.

24. Ion Pumps (page 33)

1. ATP (directly or indirectly) supplies the energy to move substances against their concentration gradient.
2. Some molecules cannot diffuse, or diffuse only slowly, through the plasma membrane so the rate of entry into the cell is insufficient to meet needs.
3. Extracellular accumulation of Na^+ (any two of):
   – maintains the gradient that is used to cotransport useful

ISBN: 978-1-927173-94-7

molecules, such as glucose, into cells.
- maintains cell volume by creating an osmotic gradient that drives the absorption of water
- establishes and maintains resting potential in nerve and muscle cells
- provides the driving force for several facilitated membrane transport proteins.

4. The accumulation of H^+ ions on one side of a membrane produces a concentration gradient. The movement of H^+ ions back across the membrane down their concentration gradient can be coupled to the otherwise energetically unfavourable transport of other molecules.

25. Excocytosis and Endocytosis (page 34)

1. Phagocytosis is the engulfment of solid material by endocytosis whereas pinocytosis is the uptake of liquids or fine suspensions by endocytosis.
2. Phagocytosis examples (any of): • Feeding in *Amoeba* by engulfing material using cytoplasmic extensions called pseudopodia. • Ingestion of old red blood cells by Küpffer cells in the liver. • Ingestion of bacteria and cell debris by phagocytic white blood cells.
3. Exocytosis examples: • Secretion of substances from specialized secretory cells in multicellular organisms, e.g. hormones from endocrine cells, digestive secretions from exocrine cells. • Expulsion of wastes from unicellular organisms, e.g. *Paramecium* and *Amoeba* expelling residues from food vacuoles.
4. (a) Oxygen: Diffusion.
   (b) Cellular debris: Phagocytosis.
   (c) Water: Osmosis.
   (d) Glucose: Facilitated diffusion.

26. Active and Passive Transport Summary (page 35)

1. A. Diffusion
   B. Osmosis
   C. Facilitated diffusion
   D. Ion pump (or sodium-potassium pump)
   E. Pinocytosis
   F. Exocytosis
   G. Phagocytosis
2. Passive transport requires no energy input from the cell (materials move down a concentration gradient). Active transport requires energy (ATP) to move substances against their concentration gradient.
3. Gases moving by diffusion: Oxygen, carbon dioxide.
4. (a) Pinocytosis
   (b) Phagocytosis.
   (c) Osmosis
   (d) Exocytosis
   (e) Endocytosis
   (f) Phagocytosis
   (g) Facilitated diffusion
   (h) Carrier-mediated facilitated diffusion
   (i) Sodium-potassium pump

27. Investigating the Origin of Life (page 36)

1. If spontaneous generation occurred then both flasks should have turned dark as they were under the same initial conditions. The swan neck to trap microbes and dust should have made no difference. Only the flask directly open to the air turned dark so the microbes must have come from somewhere, not just appeared spontaneously in the flask.
2. (a) Primeval atmosphere: Reaction chamber.
   (b) Primeval ocean: Flask containing boiling water.
   (c) Lightning: Electric discharge from power supply (7500 V, 30 A) to two tungsten electrodes.
   (d) Volcanic heat: Heating element below the flask containing the water.

28. The First Cells (page 37)

1. Ribozymes are able to act both as genes and enzymes. This enables a plausible model for the origin of life to be constructed because the ribozyme (RNA) molecules could perform the catalytic activity necessary to assemble themselves. Once formed, they could synthesize further proteins. Without a molecule (ribozyme) with this dual ability, genes could not be formed without enzymes and enzymes (proteins) could not be formed without genes.
2. Mutations in the RNA template that could provide some advantage over another RNA template (e.g. by making replication faster or producing proteins that helped in copying), would create competition for resources. An RNA able to collect the most resources would become the more predominant.

29. The Origin of Eukaryotes (page 38)

1. Sequence 1: A cell with a symbiotic "pre-mitochondrion" led to present-day animal cells. Such a cell then also engulfed a photosynthetic prokaryote (pre-chloroplast), leading to present-day plant cells.

   Sequence 2: A plant-like cell containing both aerobic and photosynthetic symbionts could have later lost its chloroplast symbiont, giving rise to animal-like cells.
2. (a) Mitochondria: Originally an endosymbiosis between an aerobic (purple) bacterium and an early cell. The bacterium was engulfed by a nucleated pre-eukaryote and became an "pre-mitochondrion". Note: The symbiosis would be progressive, with the aerobic bacterium relying more and more on the host cell for materials other than energy. The host cell would rely increasingly on the pre-mitochondrion for energy.
   (b) Chloroplasts: Originally an endosymbiosis between a photosynthetic bacterium and an early cell. The bacterium (similar to cyanobacteria) was engulfed by a nucleated pre-eukaryote and became an "pre-chloroplast". Note: As above, the symbiosis would become progressive. The host cell may or may not have already had a pre-mitochondrial structure.
3. Both mitochondria and chloroplasts retain their own self-replicating circular chromosome (DNA evidence of a prokaryotic origin). The DNA code is identical to that in prokaryotes. Chloroplasts have an internal membrane structure nearly identical to that of modern cyanobacteria.

30. The Common Ancestry of Life (page 39)

1. Molecular phylogenetics has allowed the genetic relatedness of organisms to be clarified by comparing DNA, RNA, and proteins. Similarities in morphology are not always the result of common ancestry.
2. Evidence includes archaean RNA polymerase closely resembles that of eukaryotes and the protein components of archaeal ribosomes are similar that those of eukaryotes.
3. The evidence for LUCA comes from the commonality in the genetic code and from the similarities in the molecular machinery of all cells.

31. Why Cells Need to Divide (page 40)

1. (a) Mitosis is responsible for growth in an embryo by increasing the number of cells. After birth, mitosis continues to be important in growth until the adult size is reached.
   (b) Mitosis replaces old or damaged cells with new ones (e.g. grazed skin). Other examples include the production of new cells to repair a broken bone or regeneration of a new limb in some organisms.
   (c) Some unicellular eukaryotes (and some multicellular organisms too) reproduce asexually by mitosis. For example, baker's yeast reproduce by budding, producing a genetically identical daughter cell.
2. 24

ISBN: 978-1-927173-94-7

32. Mitosis and the Cell Cycle (page 41)

1. The function of mitosis is growth and repair of the body. It may also allow asexual reproduction in some organisms.
2. The DNA must replicate (during S phase).
3. (a) Interphase: The stage between cell divisions (mitoses). Just before mitosis, the DNA is replicated to form an extra copy of each chromosome (still part of the same chromosome as an extra chromatid).
   (b) Mitosis: Homologous chromatids separate. The nucleus divides in two, each new nucleus containing a set of chromosomes.
   (c) Cytokinesis: The cytoplasm divides in two, forming two new cells.
4. Cytokinesis in animal cells involves the formation of a contractile ring of microtubules that constrict to cleave the cell. In plant cell cytokinesis, vesicles deliver cell wall a material to the middle of the cell where a cell plate (a precursor to the new cell wall) forms. The vesicles coalesce to form the plasma membranes of the new cell surfaces.

33. Recognizing Stages in Mitosis (page 43)

1. (a) Anaphase
   (b) Prophase
   (c) Metaphase
   (d) Telophase
2. (a) Estimate below. This calculation is based on seeing enough of the cell's nucleus. Cells in cytokinesis were included in interphase (not in mitosis).

Stage	No. of cells	% of total cells	Estimated time in stage
Interphase	47	41.5	9 h 57 minutes
Mitosis	66	58.6	14 h 03 minutes
Total	113	100	24 hours

   (b) 66/113 = 0.58
3. The mitotic index will decrease.

34. Regulation of the Cell Cycle (page 44)

1. If the cell cycle was not regulated cells would go into uncontrolled cell division, resulting in the formation of tumours. Cycle regulation also ensures that the cell reaches the appropriate size and volume to divide successfully.
2. (a) The cell cycle is shorter in intestinal epithelial cells as these tend to wear out, get brushed away by the movement of food through the gut, or be damaged by digestive enzymes and thus need to be constantly replaced.
   (b) The cell cycle may be shortened in response to damage, such as a cut or wound. Cells divide rapidly to heal the damage, then return to the normal rate of cell division.

35. Cancer: Cell Out of Control (page 45)

1. (a) Proto-oncogenes control the start of cell division. Tumour suppressor genes switch off cell division. The action of these two genes therefore regulates the timing of cell division.
   (b) Oncogenes remain switched on an continually signal a cell to divide. More cells than are needed are produced and a tumour forms.
2. There is a strong positive correlation between incidence of lung cancer and smoking rates. The number of lung cancer deaths per 100,00 people rises as cigarette consumption (cigarettes smoked per person per year) rises, with a lag of about 20 years.

36. Chapter Review (page 47)

No model answer. Summary is the student's own.

37. KEY TERMS: Did You Get It? (page 48)

1. active transport (F), diffusion (C), endocytosis (G), ion pump (D), mitosis (E), organelle (I), osmosis (H), passive transport (B), plasma membrane (A)
2. (a) Mitochondria (c) Chloroplast
   (b) Plant and animal cell (d) Plant cell
3 (a) Cells are the basic units of life. A cell is enclosed by a plasma membrane made of a phospholipid bilayer. A phospholipid is made up of a hydrophilic head and a hydrophobic tail. Proteins are embedded in the plasma membrane. Eukaryotic cells contain many different types of organelle some of which are composed of membranes. Each organelle carries out a specific function in the cell such as photosynthesis or respiration.
   (b) Transport of molecules through the plasma membrane can be active or passive. Active transport requires energy input whereas passive transport does not. Passive transport involves the movement of molecules from high to low concentration (down a concentration gradient). Simple diffusion can occur directly across the membrane. Facilitated diffusion involves proteins in the plasma membrane, which help molecules or ions to move through. Active transport involves membrane proteins, which couple the energy provided by ATP to the movement of molecules or ions against their concentration gradients.

38. Organic Molecules (page 51)

1. Arrows should indicate the four electrons in the outer shell.
2. In living organisms, urea is produced from other organic compounds in reactions catalysed by enzymes. Wöhler synthesized urea accidently from two inorganic compounds (originally silver cyanate with ammonium chloride) when trying to produce ammonium cyanate. Although an interesting result, the reaction did not immediately discredit vitalism as scientists interpreted the result in the context of vitalism rather than as being different to it. It took many other experiments and lines of evidence to discredit vitalism, one being the synthesis of the first true organic compound (ethanoic or acetic acid) by Adolph Kolbe in 1845.

39. Water (page 52)

1. δ symbol omitted for clarity

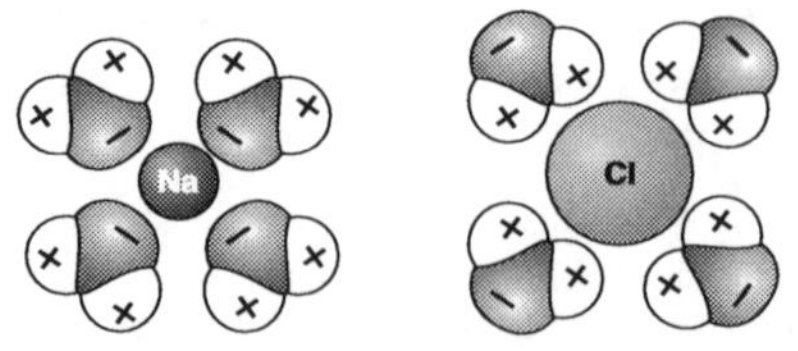

Water surrounding a positive ion (Na^+)   Water surrounding a negative ion (Cl^-)

2. Hydrogen bonds form between hydrogen and a strongly electronegative element such as oxygen. In water, hydrogen bonding helps to hold the water molecules together, making it a stable substance. A relatively large amount of energy is needed to overcome these bonds and separate the water molecules (boiling). A similar bonding occurs between water and other molecules, which makes water a very good solvent.
3. Methane molecules do not form hydrogen bonds and they are only very weakly held together. Very little energy is needed to separate molecules resulting in methane's very low melting and boiling points.

40. The Properties of Water (page 53)

1. (a) A hydrophobic molecule is not attracted to water whereas a hydrophilic molecule is attracted to water.

ISBN: 978-1-927173-94-7

(b) Hydrophilic molecules attract water. Hydrogen or ionic-dipole bonds keep the hydrophillc molecule or ion surrounded by water molecules and keep it dissolved. It thus travels through the blood as a dissolved molecule or ion, e.g. Na^+ or glucose. Hydrophobic molecules will not form intermolecular bonds with water and thus will not dissolve. They must be transported around the body by other methods, often attaching to transport molecules in cells, such as oxygen binding to hemoglobin in the blood.

2. Water absorbs heat energy from the body. The energy breaks the hydrogen bonds between the water molecules and they evaporate. Thus the body feels cooler as heat is transferred from the body to the water in the sweat.

41. Sugars (page 54)

1. (a) Primary energy source for cellular metabolism
   (b) Structural units for disaccharides and polysaccharides (energy sources and structural carbohydrates).
2. Structural isomers have the same chemical formula (number and types of atoms) but their atoms are arranged in a different sequence e.g. α and β glucose molecules.
3. Glucose provides energy for cellular respiration but is also stored as glycogen or starch for later use. Glucose can also be used to produce structural molecules such as cellulose.

42. Condensation and Hydrolysis of Sugars (page 55)

1. Disaccharide sugars are formed by condensation reactions and broken down by hydrolysis. Condensation reactions join two monosaccharide molecules by a glycosidic bond with the release of a water molecule. Hydrolysis reactions use water to split a disaccharide molecule into two. The water molecule provides a hydrogen atom and a hydroxyl group.
2. A - Condensation, product: maltose
   B - Hydrolysis, products: two glucose molecules
3.

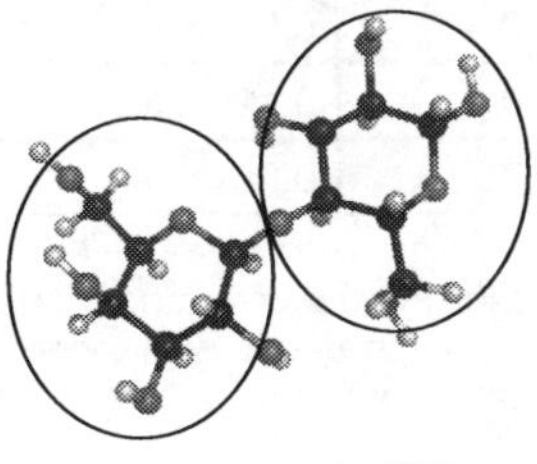

Lactose

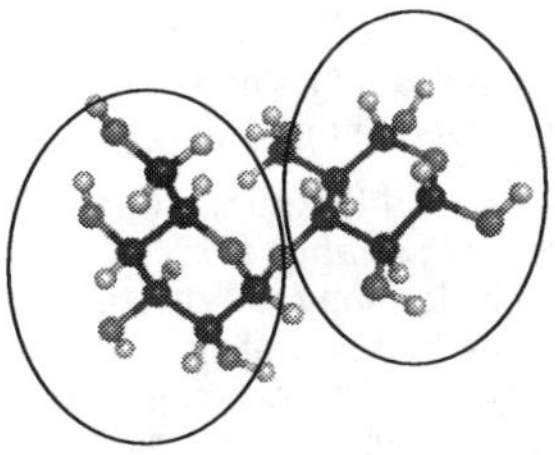

Maltose

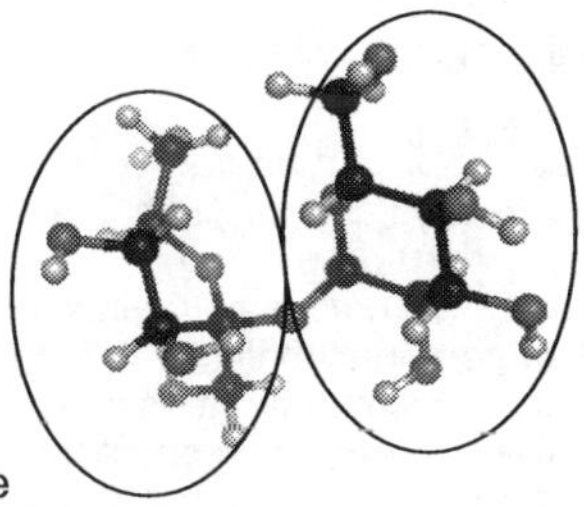

Sucrose

43. Polysaccharides (page 56)

1. (a) Polysaccharides are a good source of energy because they are easily hydrolysed into monosaccharides (e.g. glucose) when energy is needed. Monosaccharides are the primary source of cellular fuel.
   (b) Polysaccharides are hydrolysed to produce simpler carbohydrates, e.g. glucose, which can then be transported to other parts of the organism.
2. Glycogen is a highly branched glucose polymer. Its branching makes it compact, more soluble in water than starch, and easily hydrolysed to provide glucose for cellular fuel. Glycogen's properties allow it to be metabolized more quickly, which suits the active lives of moving animals. Starch is a mix of branched and unbranched chains of glucose, which makes it powdery. It is less compact than glycogen, relatively insoluble in cold water, but relatively easy to hydrolyse to soluble sugars, making it a good storage molecule for plants. Cellulose is a linear glucose polymer and is strong and insoluble, which makes it well suited to its role in providing strength and support to plant cells.

44. Starch and Glucose (page 57)

1. (a) Amyloplasts (b) The cell wall
2. Amylose is a linear molecule formed from glucose molecules held together by α-1, 4 glycosidic bonds. Amylopectin is similar but it also contains α-1, 6 glycosidic bonds, which provide branching points every 20-30 glucose monomers.
3. Cellulose is a linear molecule consisting of several hundred to several thousand β-glucose molecules bonded by a β -1,4 glycosidic bond. Starch is composed of two main molecules; amylose which forms a helix, and a branched molecule, amylopectin.
4. There are more α-1,6 glycosidic bonds in glycogen.

45. Lipids (page 58)

1. Fats need more oxidation per gram to form CO_2 than other molecules and so more usable energy can be extracted.
2. Hydrolysis of neutral fats produces free fatty acids, which undergo β-oxidation to release 2C units (as acetyl-CoA), which can enter the Krebs cycle to be respired.
3. Fatty acids derived from stored fats undergo β-oxidation in the mitochondrial matrix to produce 2C acetyl CoA units, which enter the Krebs cycle and are fully oxidized to CO_2 and water.
4. (a) Saturated fatty acids contain the maximum number of hydrogen atoms, whereas unsaturated fatty acids contain some double-bonded carbon atoms.
   (b) Saturated fatty acids tend to produce lipids that are solid at room temperature, whereas lipids that contain a high proportion of unsaturated fatty acids tend to be liquid at room temperature.
   (c) The cellular membranes of an Arctic fish contain abundant unsaturated fatty acids as this helps them to remain fluid at low temperatures.
5. (a) In *cis*- isomers, the hydrogens are on the same side of the double bond and the molecule is bent. In *trans*- isomers, the hydrogens are on the opposite site of the double bond and the molecule remains linear. The bend in *cis*- fatty acids means they don't pack together tightly and so have a lower melting point than the *trans* fats, which pack together tightly.
   (b) *Trans*- fatty acids have no specific metabolic functions (and also compete for metabolic pathways with essential fatty acids). Their structural differences mean they are metabolized differently to *cis*- isomers in the liver, elevating levels of LDL ('bad cholesterol') in the blood and increasing the risk of coronary artery disease.

46. Lipids and Health (page 60)

1. Reduction of trans-fats in the diet reduced the risk of heart disease. Using polyunsaturated fat instead of trans-fat produces the largest decrease in risk. Using monounsaturated fat or unhydrogenated, unsaturated fat instead of trans-fats produces almost the same reduction in risk.
2. (a) The differences may be based on the cohort used, e.g. there may be differences in lifestyles. However large cohorts should reduce these differences between studies. Another variable may be the technique used for gathering the information.

ISBN: 978-1-927173-94-7

(b) Health authorities agree on a lower fat diet based emphasizing a reduction in highly processed foods.

3. (a) 63 kg (b) 85.5 kg - 102.6 kg

47. Amino Acids (page 61)

1. (a) The R group
   (b) The sequence of amino acids in the polypeptide chain.
   (c) The order of nucleotides in DNA and RNA.
   (d) The differences between amino acids are due to the different properties of the R group. These cause different kinds of intermolecular bonding between amino acids in the polypeptide chain, which influences the way the polypeptide will fold up.

2 (a) A peptide bond
  (b) the bond forms by a condensation reaction

  (c)

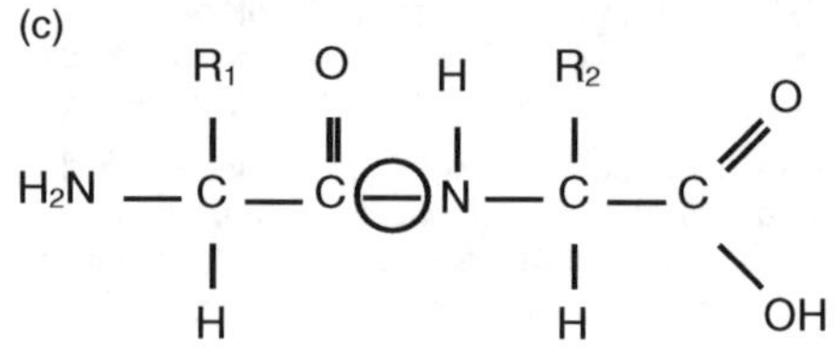

  (d) Di- and polypeptides are broken down by hydrolysis.

48. Proteins (page 62)

1. The sequence of amino acids (primary structure) determines how a protein folds. The distribution of attractive and repulsive charges on the amino acids determines how the protein is organized and folded (and therefore also determines its biological function).

2. The interior of a plasma membrane is a hydrophobic environment. Channel proteins span the membrane, and fold in such a way that the non-polar (hydrophobic) R-groups align to the outside, and polar (hydrophilic) R-groups form a channel on the inside. This channel allows water soluble molecules to cross the membrane.

3. Denaturation is the process of the protein losing it shape. The loss of shape disrupts the protein's active site and thus its ability to carry out its biological function.

49. The Role of Proteins (page 63)

1. (a) Fibrous proteins, such as collagen and elastin, are part of connective tissues, such as tendons and ligaments (and also in skin), which provide support and rigidity to more fluid components of tissues.
   (b) Enzymes are involved in almost all metabolic reactions. Examples include RuBisCo (photosynthesis), lipase (fat digestion), pyruvate dehydrogenase (cellular respiration).

2. Their tertiary structure produces long fibres or sheets, with many cross-linkages. This makes them very tough physically and ideal as structural molecules.

3. Their 3° structure produces a globular shape, which is critical to their interaction with other molecules, e.g. the active site in enzymes.

50. Enzymes (page 64)

1. (a) The active site is the region where substrate is drawn in and positioned in such a way as to promote the reaction. The properties of the active site are a function of the precise configuration of the amino acid side chains which interact with the substrate.
   (b) The active site is very specific because of how the protein folds up (its tertiary structure). It will normally accept only one type of molecule (the substrate, which has the correct configuration to interact with the active site).

2. Substrate molecules must collide with the active site.

3. Enzymes are involved in a wide variety of metabolic processes including production of ATP and NADPH in photosynthesis, the fixation of carbon from CO_2 in the Calvin cycle, glucose catalysis, and ATP production.

51. Enzyme Reaction Rates (page 65)

1. (a) Increased enzyme concentration increases reaction rate.
   (b) A cell may vary the amount of enzyme present by manufacturing more (increasing rate of synthesis).

2. (a) An increase in substrate concentration increases reaction rate initially, but the reaction rate plateaus as the amount of substrate continues to increase.
   (b) The rate changes (levels off) because, after a certain concentration of substrate, the enzymes are saturated by the substrate and the reaction rate cannot increase.

3. (a) An optimum temperature for an enzyme is the temperature where enzyme activity is maximal.
   (b) Most enzymes perform poorly at low temperatures because chemical reactions occur slowly or not at all at low temperatures (enzyme activity will reappear when the temperature increases; usually enzymes are not damaged by low temperatures).

4. (a) Optimum pH: pepsin: 1-2, trypsin: approx. 7.5-8.2, urease: approx. 6.5-7.0.
   (b) The stomach is an acidic environment which is the ideal pH for pepsin.

52. Investigating Catalase Activity (page 66)

1. $2H_2O_2 \rightarrow 2H_2O + O_2$

2. (a)-(c), mean, standard deviation, and mean rate below.

Stage	Mean	Std Dev	Mean rate ($cm^3\ s^{-1}\ g^{-1}$)
0.5	10.1	0.5	0.03
2	34.9	3.8	0.12
4	65.5	5.0	0.22
6	36.7	4.0	0.12
10	22.5	2.7	0.08

3. (a) The values obtained for the 0.5 days and 10 days of germination are not in accordance with those obtained by the other groups (trials). The other values are of a similar magnitude.
   (b) If the new data are used, the mean should exclude those two values and be calculated using only 5 (rather than 6) trials for those germination stages.
   (c) These data values are clearly well adrift from the other values obtained for those stages from other groups (trial); if plotted as a scatter plot, they are distinct outliers on the plot. This suggests that something was wrong with either the measurement or the execution of the trial. It is reasonable therefore to exclude them from the analysis.

4. See the graph at the top of the next column

5. (a) The volume and rate of oxygen production increases rapidly to a peak at 4 days and declines, almost as sharply between 4 and 10 days.
   (b) Catalase activity in the sprouting seeds increases rapidly in the first 4 days of germination linked to the increase in cell activity and high respiration rates in early growth. It then falls off as the seedlings become established and metabolism slows.

6. Errors include: The equipment could leak around the bung or the tubing. There could be a delay in delivering all the H_2O_2 so a slight delay in correctly timing the start of the reaction. The seeds might not be completely crushed, or crushed to different degrees so that not all the catalase is released.

7. Validity of data could be affected by (two of): insufficient usable data, highly variable data (overlapping data between times), old or poorly stored beans, old or poorly stored H_2O_2, precise, reliable but inaccurate data because of gas losses

ISBN: 978-1-927173-94-7

through the equipment.

Catalase activity in relation to stage of germination in mung bean seedlings

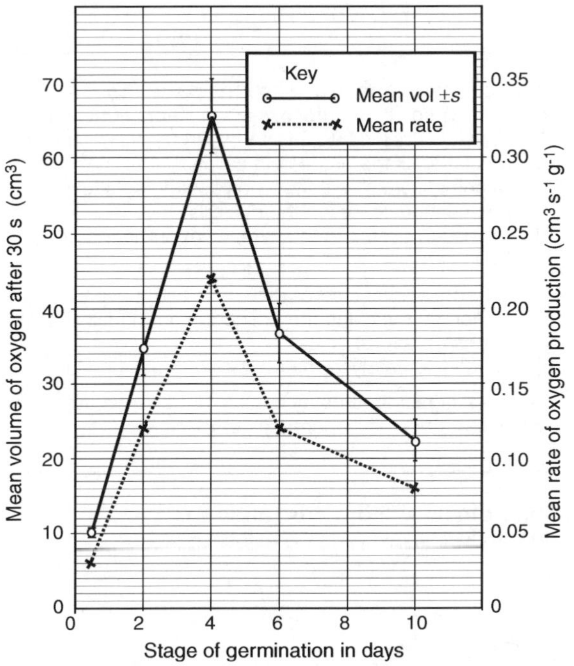

8. Improvements could include any one of: Using a mechanical grinder for a set period of time to ensure complete release of catalase. Using a petroleum sealant to prevent air leaks. Making sure the same person takes the measurement in each case, making sure all the trials are carried out with the same equipment by the same team.

53. Applications of Enzymes (page 68)

1. Milk is a very good food source (it contains good levels of macronutrients, vitamins, and minerals). Being able to drink milk throughout life means all people could benefit from its nutritional value.
2. The enzyme lactase is immobilized to remove lactose from milk. It does so by binding the lactose and enzymatically hydrolysing it into glucose and galactose.
3. One of the products of lactose hydrolysis (glucose) is sweeter than lactose itself. Therefore, lactose free milk tastes sweeter than ordinary milk.

54. Nucleotides and Nucleic Acids (page 69)

1. Labels as follows (only half of the section of DNA illustrated in the workbook is shown):

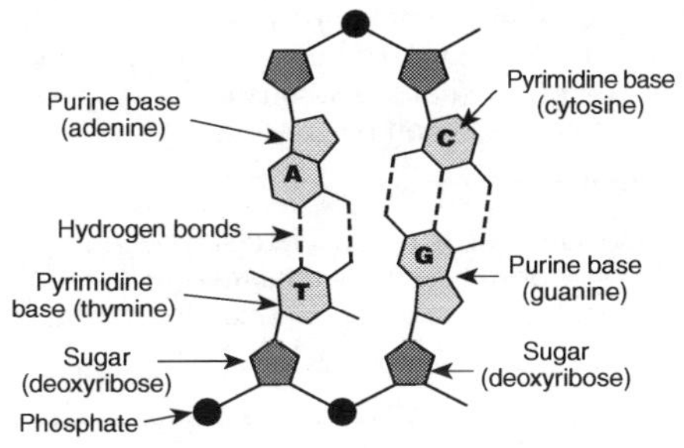

2. (a) The following bases always pair in a normal DS DNA: guanine with cytosine, cytosine with guanine, thymine with adenine, adenine with thymine.
   (b) In mRNA, uracil pairs with adenine (not thymine).
   (c) The hydrogen bonds in double stranded DNA hold the two DNA strands together.
3. **Nucleotides** are building blocks of nucleic acids (DNA, RNA). Their precise sequence provides the genetic blueprint for the organism.
4. (a) The asymmetric (phosphodiester) bonds in the sugar-phosphate backbone give the molecule a direction so that the two strands in the double-helix run in opposite directions (they are anti-parallel).
   (b) 5' end terminates in a phosphate group, 3' end terminates in a hydroxyl group (from a sugar).
5.

	DNA	RNA
Sugar present	Deoxyribose	Ribose
Bases present	Adenine	Adenine
	Guanine	Guanine
	Cytosine	Cytosine
	Thymine	Uracil
Number of strands	Two (double)	One (single)
Relative length	Long	Short

55. Creating a DNA Model (page 71)

3. Labels as follows:

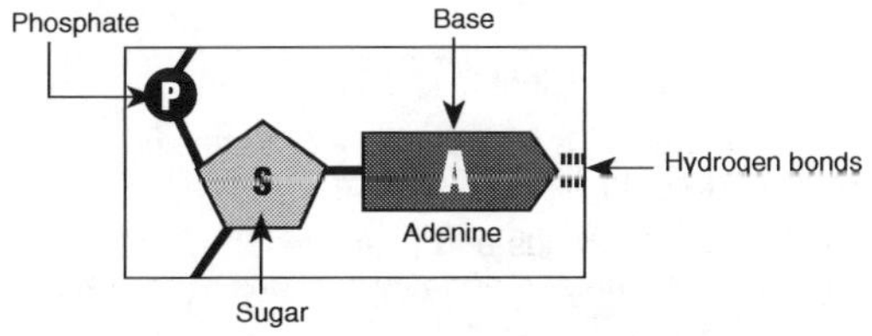

4.and 5. below.

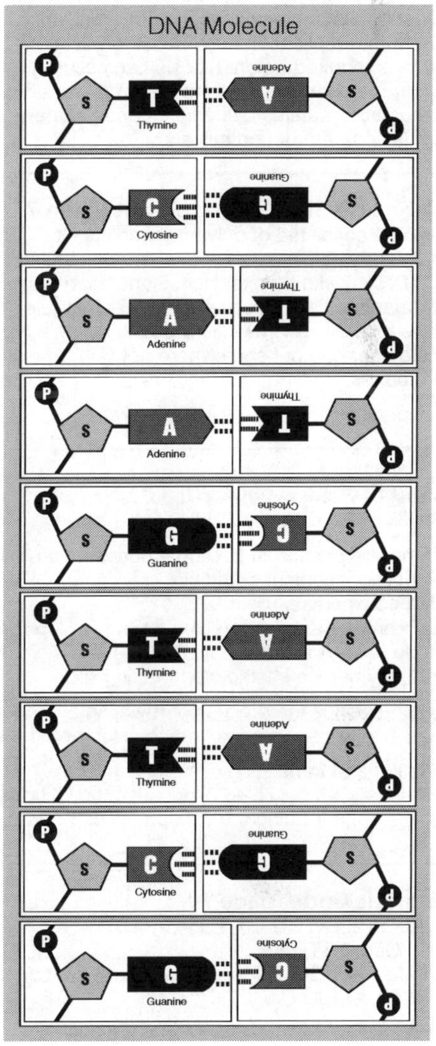

ISBN: 978-1-927173-94-7

6. Factors preventing mismatch of nucleotides:
   - The number of hydrogen bond attraction points.
   - The size (length) of the base (thymine and cytosine are short, adenine and guanine are long).
     Examples: Cytosine will not match cytosine because the bases are too far apart, G will not match G because they are too long to fit side by side; T will not match G and C will not match A because there is a mis-match in the number and orientation of H-bonds.

56. DNA Replication (page 75)

1. DNA replication prepares a chromosome for cell division by producing two chromatids which are identical copies of the genetic information for the chromosome.
2. (a) Step 1: Enzymes unwind DNA molecule to expose the two original strands.
   (b) Step 2: DNA polymerase enzyme uses the two original strands as template to make complementary strands.
   (c) Step 3: The two resulting double-helix molecules coil up to form two chromatids.
3. 44
4. 50% new 50% original
5. It means that each new double strand has one strand that is original (parent) DNA.
6. (a) Helicase unwinds and separates the DNA strands.
   (b) DNA polymerase adds nucleotides to the new DNA strand in the 5' to 3' direction.
7. DNA replication is the process by which the DNA molecule is copied to produce two identical DNA strands. Replication is tightly controlled by enzymes. The enzymes also proofread the DNA during replication to correct any mistakes. DNA replication is required before mitosis can occur. After replication, the chromosome is made up of two chromatids. Each chromatid contains half original and half new DNA. The chromatids separate during mitosis.

57. Meselson and Stahl's Experiment (page 77)

1. Generation 1 consisted of only intermediate DNA. This ruled out the conservative method where equal quantities of heavy and light DNA would be expected. Generation 2 produced light and intermediate DNA in equal amounts. This ruled out the dispersive method which would produce a single type of DNA that would separate out somewhere between the intermediate and light bands.
2. (a) Conservative (b) Dispersive

58. Genes to Proteins (page 78)

1 (a) A triplet (b) A codon

2 (a) A gene is a section of DNA that codes for a protein. it is the functional unit of heredity.
  (b) RNA polymerase.
  (c) The promoter acts as a region for the RNA polymerase to bind to to start transcription. A terminator indicates where RNA polymerase should stop transcribing.

3. Gene expression is the process of rewriting a gene into a protein. It involves two stages, transcription and translation.
4. Transcription and translation occur only in 5' to 3' direction (polymerases work only 5' to 3' because the OH on the 3' is the active site for forming the phosphodiester bond).

59. The Genetic Code (page 79)

1. (a) DNA: ATG GGT TAC CTG AGG GTA ATA CGG GCA CTT TAG
   (b) mRNA: AUG GGU UAC CUG AGG GUA AUA CGG GCA CUU UAG
   Amino acids: Met Gly Tyr Leu Arg Val Iso Arg Ala Leu STOP
2. (a) Start: AUG
   STOP: UAA, UAG, UGA
   (b) Any example where more than one codon codes for the same amino acid, e.g. CGU, CGC, CGA, CGG, AGA, and AGG all code for arginine.

60. Transcription and Translation (page 80)

1. (a) Transcription is the process in which the template strand of DNA is copied into mRNA by RNA polymerase.
   (b) Translation is the process in which the mRNA is read by ribosomes, which catalyse polypeptide synthesis.
   (c) Transcription occurs in the nucleus.
   (d) Translation occurs in the cytoplasm on bound or free ribosomes.
2. (a) tRNA molecules bind amino acids and move them into position by matching the anticodon on the tRNA with codons on the mRNA.
   (b) Ribosomes catalyse the synthesis of polypeptides.
3. The advantage is that many copies of the same protein can be made at once. This increases efficiency.

61. The Role of ATP in Cells (page 81)

1. ATP production occurs in the mitochondrion.
2. ATPase
3. (a) and (b)

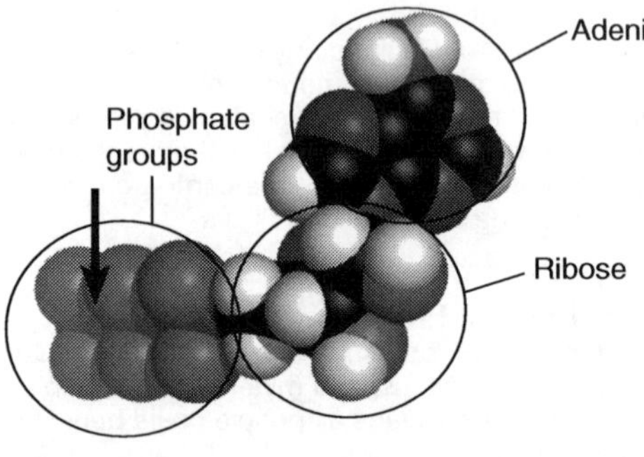

4. (a) ATP is required for the muscular activity involved in shivering (used to heat the body when it is cool). ATP is also required for the glandular secretion involved in sweating (used to cool the body).
   (b) ATP hydrolysis provides the energy needed to move flagella in motile cells (e.g. a sperm cell).
   (c) ATP is needed to provide energy for movement of filaments in the cytoskeleton, production of the nuclear membrane, and cytokinesis.
5. (a) In the presence of the enzyme ATPase, ATP is hydrolysed to produce ADP plus a free phosphate, releasing energy in the process.
   (b) Like a rechargeable battery, the ADP/ATP system alternates between a high energy and a low energy state. The addition of a phosphate to ADP recharges the molecule so that it can be used for cellular work.
6. Glucose (or pyruvate)
7. Highly active cells (sperm, muscle cells) need a lot of energy (as ATP) to perform their functional role (movement). ATP production occurs in the mitochondria, so they have large numbers to generate enough ATP to meet energy demands.

62. Cell Respiration (page 83)

1. The purpose of cellular respiration is to produce ATP.
2. (a) Glycolysis: cytoplasm
   (b) Krebs cycle: matrix of mitochondria
   (c) Electron transport chain: cristae (inner membrane surface) of mitochondria.

ISBN: 978-1-927173-94-7

3 For maximum theoretical yield:
(a) Glycolysis: 2 ATPs
(b) Krebs cycle: 2 ATPs
(c) Electron transport chain: 34 ATPs
(d) Total produced: 38 ATPs

63. Measuring Respiration (page 84)

1. (a) RQ at 20°C after 48 h = 1.97 ÷ 2.82 = **0.7** (0.698)
   (b) RQ at 20°C after 1 h = 2.82 ÷ 2.82 = 1. After 2 days without feeding the cricket was metabolizing fats for energy. Shortly (1 hour) after feeding it was metabolizing only carbohydrate (RQ = 1).
2. (a) RQ of two seedlings during early germination:

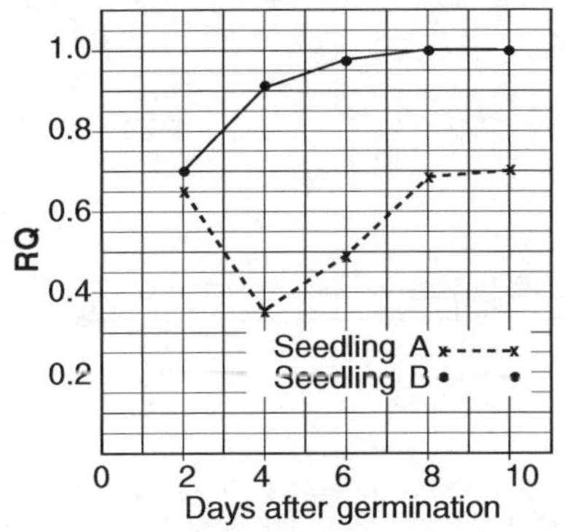

   (b) Both seedlings began their germination metabolizing primarily fats for energy. However, while seedling A continued to metabolize mainly fats (with some synthesis of carbohydrate and organic acids), throughout the 10 day period, seedling B rapidly moved to metabolizing carbohydrate alone. **Note**: The value of 0.91 (seedling B) may have been the result of protein metabolism alone or (more likely) respiration of a mix of fat and glucose.

64. Anaerobic Metabolism (page 85)

1. (a) 2 ÷ 38 x 100 = 5.3% efficiency (theoretical maximum).
   (b) Only a small amount of the energy of a glucose molecule is released in anaerobic respiration. The remainder stays locked up in the molecule.
2. The build up of toxic products (ethanol or lactate) inhibits further metabolic activity.
3. Respiration of lactate provides more energy for working muscle cells enabling greater intensity and duration of activity.

65. Investigating Alcoholic Fermentation in Yeast (page 85)

1. $C_6H_{12}O_6 \rightarrow 2C_2H_5OH + 2CO_2$
2. Rate of CO_2 production, group 1 (SI units also given):
   (a) None: 0 $cm^3\ min^{-1}$ or 0 $cm^3\ s^{-1}$
   (b) Glucose: 0.443 $cm^3\ min^{-1}$ or 7.4 x 10^{-3} $cm^3\ s^{-1}$
   (c) Maltose: 0.24 $cm^3\ min^{-1}$ or 4.0 x 10^{-3} $cm^3\ s^{-1}$
   (d) Sucrose: 0.191 $cm^3\ min^{-1}$ or 3.2 x 10^{-3} $cm^3\ s^{-1}$
   (e) Lactose: 0.029 $cm^3\ min^{-1}$ or 4.8 x 10^{-4} $cm^3\ s^{-1}$
3. Rate of CO_2 production, group 2 (SI units also given):
   (a) None: 0 $cm^3\ min^{-1}$ or 0 $cm^3\ s^{-1}$
   (b) Glucose: 0.271 $cm^3\ min^{-1}$ or 4.5 x 10^{-3} $cm^3\ s^{-1}$
   (c) Maltose: 0.211 $cm^3\ min^{-1}$ or 3.5 x 10^{-3} $cm^3\ s^{-1}$
   (d) Sucrose: 0.194 $cm^3\ min^{-1}$ or 3.2 x 10^{-3} cm3 s^{-1}
   (e) Lactose: 0 $cm^3\ min^{-1}$ or 0 $cm^3\ s^{-1}$
4. The assumptions made are that 24°C and pH 4.5 provide suitable (even optimal) conditions for yeast fermentation. This is reasonable as it has been stated in the background that the literature cites warm, slightly acidic conditions as optimal for baker's yeast.
5. Graph of Group 1's results: see top of next page.
6. (a) The fermentation rates were greatest for glucose, with a CO_2 yield approximately twice that for maltose and sucrose. Maltose and sucrose were similar to each other, while the rate of fermentation on lactose was minimal.
   (b) Glucose is the preferred substrate (directly available as a fuel). Maltose (glucose-glucose) and sucrose (glucose-fructose) must first be hydrolysed before the glucose is available (the fructose from sucrose must also be isomerized to glucose). This accounts for the lower fermentation rates on these substrates. Lactose is a poor fuel, presumably because yeast lack the enzyme to hydrolyse its galactose and glucose components.
7. (a) Column chart comparing results of the two groups.

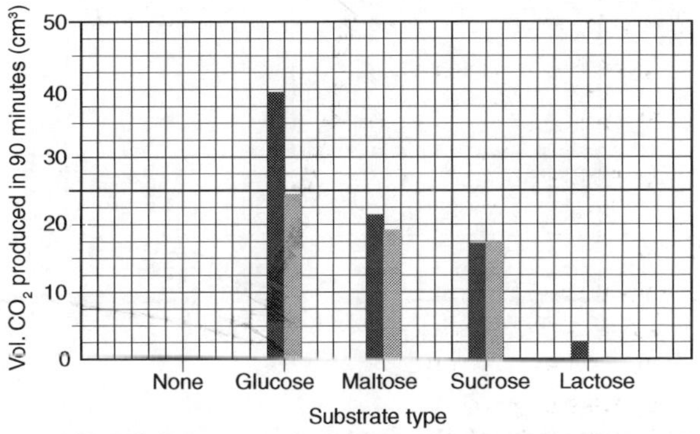

   (b) Main points to note are that both groups had results of similar magnitude for maltose and sucrose. Group 1's fermentation rates were much higher on glucose but group 2's were only a little higher on glucose than those for maltose and sucrose. Group 1 had some fermentation response for lactose, but group 2 had the same result as for no substrate.
   (c) Both groups had very active fermentation on glucose, sucrose, and maltose. Group 1's finding that yeast fermented lactose is probably erroneous as yeast lack the enzyme required to hydrolyse lactose to component monosaccharides. The fact that group 2 did not obtain any CO_2 production for lactose suggests that group 1's lactose was contaminated with some other substrate, accounting for a low level of activity. Group 1's very high rates of activity on glucose could be attributable to using a different batch of (fresher) yeast, but this would have to be tested further.
   (d) The main improvement would be to repeat the experiment 3 or 4 times under the same conditions, or run replicates, so that a mean volume and rate of CO_2 production could be calculated for each substrate. This would help to account for some of the variability. Ensuring that all tests are made with the same batch of yeast and that the yeast is still active (i.e. not old stock) would also be worthwhile. Students should also note a procedure to ensure no cross contamination between substrates.

66. Photosynthesis (page 88)

1. (a) Water (or carbon dioxide)
   (b) Carbon dioxide (or water)
   (c) Photosynthesis
   (d) Oxygen
   (e) Glucose

67. Pigments and Light Absorption (page 89)

1. The absorption spectrum of a pigment is that wavelength of the light spectrum absorbed by a pigment, e.g. chlorophyll absorbs red and blue light and appears green. Represented graphically, the absorption spectrum shows the relative amounts of light absorbed at different wavelengths.
2. Accessory pigments absorb light wavelengths that chlorophyll a cannot absorb, and they pass their energy on to chlorophyll *a*. This broadens the action spectrum over which chlorophyll a can fuel photosynthesis

ISBN: 978-1-927173-94-7

Cumulative volume of CO_2 produced over 90 minutes by yeast grown on different carbohydrate sources.

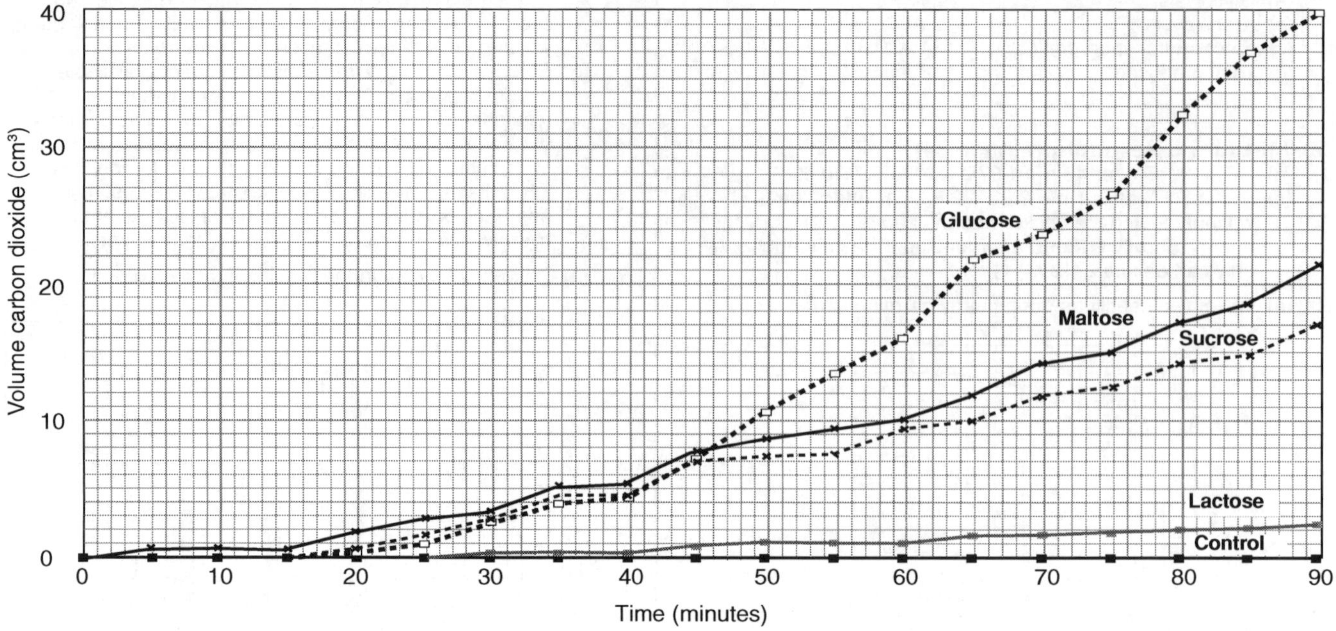

68. Separation of Pigments by Chromatography (page 90)

1. Rf = 15 mm ÷ 33 mm = 0.45
2. Rf must always be less than one because the substance cannot move further than the solvent front.
3. Immersion would just wash out the substance into solution instead of separating the components out behind a solvent front.
4. Leucine, arginine, alanine, glycine (most soluble to least soluble).
5. Chlorophyll is not as soluble in acetone as carotene or xanthophyll.

69. Factors Affecting Photosynthetic Rate (page 91)

1. (a) CO_2 concentration: Photosynthetic rate increases as the CO_2 concentration increases, and then levels off.
   (b) Light intensity: Photosynthetic rate increases rapidly as the light intensity increases and then levels off.
   (c) Temperature: Increase in temperature increases the rate of photosynthesis, but the effect is not marked at low CO_2.
2. The photosynthetic rate is determined by the rate at which CO_2 enters the leaf. When this declines because of low atmospheric levels, so does photosynthetic rate.
3. (a) By changing only one factor at a time (either temperature or CO_2 level) it is possible to assess the effects of each one separately.
   (b) CO_2 has the greatest effect of these two variables.
   (c) At low levels of CO_2, increase in temperature has little effect (the rate of CO_2 entry into the leaf is the greatest determinant of photosynthetic rate).
4. Glasshouses are controlled environments in which the levels of light, humidity, temperature, and CO_2 can be optimized to maximize photosynthetic rate. In particular, CO_2 enrichment in an environment where transpiration losses are minimised enables maximum productivity.
5. Student's own experimental design. Possible hypothesis is that temperature will increase the rate of photosynthesis within limits (H_0 being no effect of temperature). The choice of design and plant will be up to the student but an easy one to choose is to measure the rate of oxygen (air bubble) production in a submerged aquatic macrophyte such as *Elodea*. The temperature can then be easily changed by heating or cooling the water. Things the student should make reference to and explain if necessary:
   - the control and how it was set up
   - how the dependent variable (biological response) is to be measured and how many measurements will be made
   - how the independent variable (temperature) is to be varied and the range of that variation
   - how other variables are kept constant and the value of these variables (e.g. light level)
   - how many plants at each treatment level

70. Chapter Review (page 92)

No model answer. Summary is the student's own.

71. KEY TERMS: Did You Get it? (page 94)

1. amino acid (M), ATP (E), catabolism (P), catalyst (I), condensation (A), denaturation (K), disaccharide (C), DNA (R), enzyme (G), fermentation (Q), hydrolysis (B), lipid (J), monosaccharide (N), nucleic acid (O), polypeptides (L), polysaccharides (H), semi- conservative (F), translation (D)
2. (a) Cell respiration
   (b) In the mitochondria of cells
   (c) ATP
   (d) By hydrolysis of the terminal phosphate coupled to another reaction.
3. (a) Carbon dioxide + water + light → Glucose + oxygen + water
   $6CO_2 + 12H_2O + light \rightarrow C_6H_{12}O_6 + 6O_2 + 6H_2O$
   (b) The chloroplasts

72. Alleles (page 97)

1. (a) Heterozygous: Each of the homologous chromosomes contains a different allele for the gene (one dominant and one recessive).
   (b) Homozygous dominant: Each of the homologous chromosomes contains an identical dominant allele.
   (c) Homozygous recessive: Each of the homologous chromosomes contains an identical recessive allele.
2. (a) Aa (b) AA (c) aa
3. Each chromosome of a homologous pair comes from a different parent: one of maternal origin, one of paternal origin (they originated from the egg and the sperm that formed the zygote). They contain the same sequence of genes for the same traits, but the versions of the genes (alleles) on each chromosome may differ.
4. Alleles are different versions of the same gene. Different

ISBN: 978-1-927173-94-7

alleles provide phenotypic variation for the expression of a gene. There are often two alleles for a gene, one dominant and one recessive. In this case, the dominant allele will be expressed in the phenotype. Sometimes alleles for a gene can be equally dominant, in which case, both alleles will be expressed in the phenotype. Where three or more alleles for a gene exist in a population, there is more phenotypic variation in the population (for that phenotypic character) than would be the case if there were just two alleles.

73. Changes to the DNA Sequence (page 98)

1. A mutation is a change in the DNA base sequence caused by a DNA copying error.
2. A deletion mutation.
3. A person heterozygous for the mutation would not be affected (only people with the two recessive alleles are deaf).
4. (a) i Deletion
   ii Insertion
   iii Substitution

   (b) The 6th triplet codes TGA which is a stop sequence. The protein coded for would not be complete.

74. Mutagens (page 99)

1. (a) Radiation: Ultraviolet (UV) rays from the sun, X-rays from medical equipment, gamma-rays from medical equipment and nuclear contamination.
   (b) Chemicals: Benzene, formalin (formaldehyde), carbon tetrachloride.
2. A mutagen increases the frequency of mutation (change or disruption in DNA) above the spontaneous (background) rate. Mutagens bring about their effects by disrupting the base sequence of genes, which can affect that gene's product. Mutations to regulatory genes, such as those controlling cell division, are among the most damaging.
3. Somatic mutations occur in the body (non-gametic or somatic) cells and are not inherited. They may affect an individual within its lifetime. Gametic mutations are mutations to the gametes (in testes or ovaries) and are inherited. Gametic mutations are inherited. They are passed on to the next generation and can become part of the genetic variation in the gene pool (upon which natural selection can act).

75. Gene Mutations and Genetic Diseases (page 100)

1. (a) Gene name: HBB Chromosome: 11
   Mutation type: Substitution.
   (b) Gene name: CFTR Chromosome: 7
   Mutation type: autosomal recessive. Great range: deletion, missense, nonsense, misplaced terminator codon. Most common is a deletion of 3 nucleotides.
   (c) Gene name: HTT (aka IT15) Chromosome: 4
   Mutation type: autosomal dominant. Duplication (CAG repeats of varying length).
2. (a) Autosomal recessive
   (b) Autosomal dominance
   (c) Autosomal codominance
3. The disease does not become apparent until mid-adulthood by which time a sufferer has likely already had children.

76. The Sickle Cell Mutation (page 101)

1. (a) Bases: 21
   (b) Triplets: 7
   (c) Amino acids coded for: 7
2. mRNA: GUG CAC CUG ACU CCU GAG GAG
3. Amino acids: Val His Leu Thr Pro Glu Glu
4. Mutant DNA: CAC GTG GAC TGA GGA CAC CTC

   Type of mutation: Substitution
5. Mutant mRNA: GUG CAC CUG ACU CCU GUG GAG
6. Amino acids coded by mutant DNA:

   Val His Leu Thr Pro Val Glu
7. The mutation causes a change to one amino acid in hemoglobin. The mutated hemoglobin is less soluble, distorts the red blood cells, and causes severe circulatory problems.
8. The sickle cell mutation affords some degree resistance to malaria and so persists where malaria is present.

77. Genomes (page 102)

1 (a) 49.67 times bigger
  (b) 652.2 times smaller
  (c) 18,750 times smaller
2. Large genomes require a lot of resources to replicate, thus reproduction and growth rates are low and this increases the risk of extinction.
3. (a) 16,569
   (b) 248.96 MB
   (c) 42.3%

78. Prokaryotic Chromosome Structure (page 103)

1. (a)-(b) any two:
   - The prokaryotic chromosome is a singular circular chromosome. Eukaryote chromosomes comprise linear DNA packaged with proteins.
   - In prokaryotes, some genes are carried on extra-chromosomal plasmid DNA.
   - The prokaryote chromosome is attached to the plasma membrane and is not enclosed in a nuclear membrane (unlike the eukaryotic chromosomes).
   - Prokaryotic chromosomes consists almost entirely of protein coding genes and their regulatory sequences. Eukaryotic chromosomes contain much intronic DNA that does not code for proteins.
2. Transcription (making mRNA) and translation (protein synthesis) can occur at the same time because the DNA in prokaryotes is in direct contact with the cytoplasm. The entire process of gene expression occurs much more rapidly than in eukaryotic cells, where the mRNA must leave the nucleus and enter the cytoplasm before translation can begin.
3. The bacterial genome (chromosome) is much smaller than the eukaryotic genome (all chromosomes).

79. Eukaryotic Chromosome Structure (page 104)

1. In the nucleus
2. The DNA molecule is very large. Without packaging it up it would be difficult to fit the DNA into the cell nucleus, especially during replication and mitosis.
3. DNA wraps around the histone proteins, taking up less space that if it was spread out.
4. Exons are protein-coding regions of DNA. Introns are non-protein coding regions of DNA (extra note: introns are excised from the primary transcript mRNA to form the mature mRNA).

80. Stages in Meiosis (page 105)

1. Meiosis produces variation in the gametes through crossing over and recombination of alleles between homologues, and independent assortment of maternal and paternal chromosomes into the gametes.
2. Meiosis I is a reduction division in which homologous chromosomes are separated into different intermediate cells. Meiosis II is a mitotic division in which the sister chromatids are separated into separate gametes.

ISBN: 978-1-927173-94-7

81. Modelling Meiosis (page 106)

The genotype/phenotype of the offspring students obtain will depend on their own phenotypes.

82. Non-disjunction in Meiosis (page 108)

1. Non-disjunction results in abnormal gamete numbers in some gametes. As a result, certain phenotypic traits are exhibited (e.g. facial features, mental retardation).Certain metabolic processes may also be affected.
2. Non-disjunction in the parental cell during meiosis I results in both the daughter cells being faulty which will be transferred to the daughter cells produced in meiosis II. Nondisjunction in a parental cell during meiosis II results in only half the total number of daughter cells being faulty.
3. The maternal age effect refers to an increased risk of chromosome abnormalities with advancing maternal age. The risk of aneuploidies, such as Down syndrome, increase with every year a women ages.

83. Karyotypes (page 109)

1. (a) A karyogram is the physically laid out image (i.e. a picture) of the karyotype. The karyotype is the complement of a cell or organism, characterized by the number, size, shape, and centromere position of the chromosomes.
   (b) It can provide information on gender and chromosomal abnormalities.
2. Autosomes are non-sex chromosomes that occur as 'matched' homologoues pairs. Sex chromosomes (also called heterosomes) are the pair of chromosomes (XX in female humans and XY in male humans). Autosomes are not involved in determining gender, whereas the sex chromosomes are.
3. Number the chromosomes: (see below)
4. Circle the sex chromosomes (see below):

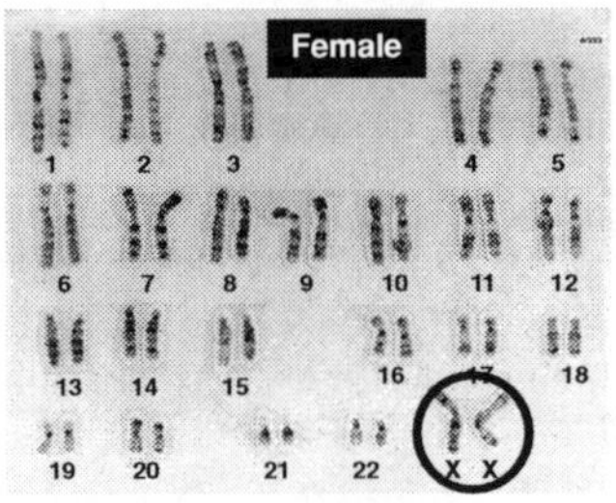

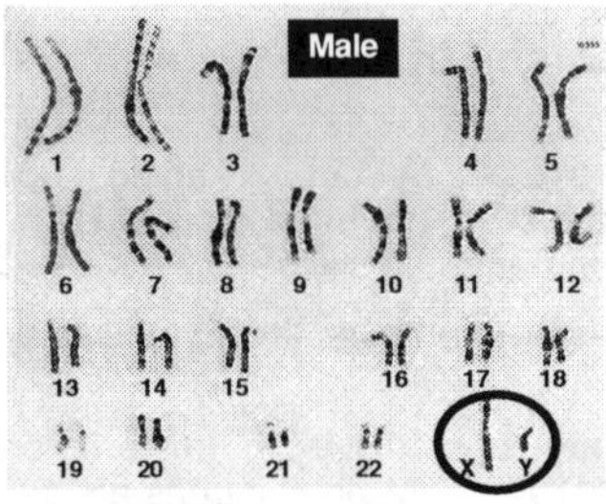

5. (a) Female autosomes: 44 Sex chromosomes: XX
   (b) Male autosomes: 44 Sex chromosomes: XY
6. (a) Chromosomes in human somatic cell: 46
   (b) Chromosomes in human sperm or egg cell: 23

84. Human Karyotype Exercise (page 111)

By studying the distinguishing characteristics of the chromosomes, you should be able to arrange the them in their correct sequence on the karyotype record sheet.

1. The karyotype should be organized as for the "Typical layout of a human karyotype" on page 143, but it has one extra chromosome number 21 (trisomy 21) making it a Down syndrome male.
2. (a) Sex: Male
   (b) Abnormal
   (c) 45 + XY (trisomy 21 or Down syndrome).

85. The Human Genome Project (page 114)

1. The HGP aimed to map the entire base sequence of every chromosome in the human cell (the human genome), to identify all genes in the sequence, determine what they express (protein produced), and determine the precise role of every gene on the chromosomes.
2. (a) 53%
   (b) 8%
   (c) Knowing the DNA sequence and position of genes will enable better screening for genetic diseases and the development of drugs or other treatments for them.
3. Chromosome 9

86. Mendel's Pea Plant Experiments (page 115)

1. and 2.:

	Dominant	Recessive	Ratio
Seed colour	Yellow	Green	3.01 : 1
Pod colour	Green	Yellow	2.82 : 1
Flower position	Axial	Terminal	3.14 : 1
Pod shape	Constricted	Inflated	2.95 : 1
Stem length	Tall	Dwarf	2.84 : 1

3. (a) Seed shape (2.96:1), seed colour (3.01:1), and pod shape (2.95:1).
   (b) Considering all the traits, larger sample sizes generally produced ratios closer to the predicted theoretical ratio. Smaller samples are more likely to produce results that deviate from the theoretical ideal, because they are affected more by the random nature of meiosis and fertilization.

87. Mendel's Laws of Inheritance (page 116)

1. Particulate inheritance: Inherited characteristics are transmitted by discrete entities (genes) which themselves remain unchanged from generation to generation.
2. (a) Aa (b) A, A, a, a (c) 2 kinds: A and a
3. Note: During meiosis, all combinations of alleles are distributed to gametes with equal probability. The pair of alleles for each gene are sorted independently of those for all other genes. Genes that are linked on the same chromosome tend to be inherited together.
   (a) AB and ab (b) 4 kinds: AB, Ab, aB, ab

88. Basic Genetic Crosses (page 117)

1. All the F_1 generation are heterozygous and thus display the dominant flower colour. When these are crossed together a 3:1 phenotypic ratio appears so that the homozygous recessive feature will be seen again.
2.

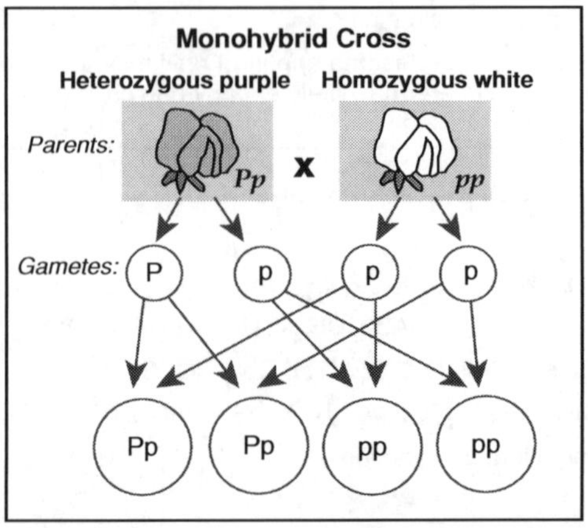

ISBN: 978-1-927173-94-7

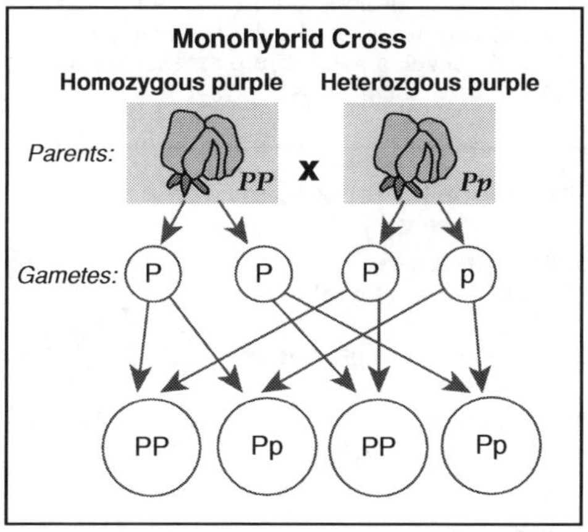

89. The Test Cross (page 118)

1. To perform the test cross here you would have to cross the male with a female (brown eye, ebony body) homozygous recessive for both genes (bbee).

   Note: Depending on the male's genotype the outcomes of the test cross would produce offspring that were:

   - All wild type (brown body red eye)
   - Half wild type, and half red eye, ebony body
   - Half wild type, and half brown eye, brown body
   - 25% wild type, 25% red eye ebony, 25% brown eye, brown body, and 25% brown eye, ebony.

   Students could present this in Punnett squares.

Male gametes

Female gametes

	BE
eb	BbEe

red eye, brown body

	BE	Be
eb	BbEe	Bbee

red eye, brown body; red eye, ebony body

	BE	bE
eb	BbEe	bbEe

red eye, brown body; brown eye, brown body

	BE	Be	bE	be
eb	BbEe	Bbee	bbEe	bbee

red eye, brown body; red eye, ebony body; brown eye, brown body; brown eye, ebony body

2. A wild type male (brown or normal body and red eyes) could have one of four genotypes: BEBE, BEbE, BEBe, or BEbe
3. (a) 50% are wild type and 50% are red eye, ebony so the male must BEBe.
   (b) This is the only genotype that produces this ratio of phenotypes in the offspring (as per the second Punnett square above)

90. Monohybrid Cross (page 119)

1.

	Genotype	Phenotype
Cross 2	50% BB 50% Bb	100% black
Cross 3	25% BB 50% Bb 25% bb	75% black 25% white
Cross 4	100% BB	100% black
Cross 5	50% Bb 50% bb	50% black 50% white
Cross 6	100% bb	100% white

91. Codominance of Alleles (page 120)

1. Two or more alleles are dominant over any recessive alleles and both are fully expressed.
2. (a) Diagram labels:

	White bull	Roan cow
Parent genotype:	C^WC^W	C^RC^W
Gametes:	C^W, C^W	C^R, C^W
Calf genotypes:	C^RC^W, C^WC^W	C^RC^W, C^WC^W
Phenotypes:	roan, white	roan, white

   (b) Phenotype ratio: 50% roan, 50% white (1:1)
   (c) By breeding only from the roan calves. Offspring of roan parents should include white, roan, and red phenotypes. By selecting only the red offspring from this generation it would be possible to breed a pure herd of red cattle.
3. (a) Diagram labels:

	Unknown bull	Roan cow
Parent genotype:	C^RC^R	C^RC^W
Gametes:	C^R, C^R	C^R, C^W
Calf genotypes:	C^RC^R, C^RC^W	C^RC^R, C^RC^W
Phenotypes:	red, roan	red, roan

   (b) Unknown bull: red bull
4. The phenotypic ratio would be 1 red: 2 roan: 1 white

92. Codominance in Multiple Allele Systems (page 121)

1. Blood group table:

Blood group B	I^BI^B, I^Bi
Blood group AB	I^AI^B

2.

Cross 2	**Group O**	**Group O**
Gametes:	i, i	i, i
Children's genotypes:	ii, ii	ii, ii
Blood groups:	O, O	O, O
Cross 3	**Group AB**	**Group A**
Gametes:	I^A, I^B	I^A, i
Children's genotypes:	I^AI^A, I^Ai	I^BI^A, I^Bi
Blood groups:	A, A	AB, B
Cross 4	**Group A**	**Group B**
Gametes:	I^A, I^A	I^B, i
Children's genotypes:	I^AI^B, I^Ai	I^AI^B, I^Ai
Blood groups:	AB, A	AB, A
Cross 5	**Group A**	**Group O**
Gametes:	I^A, i	i, i
Children's genotypes:	I^Ai, I^Ai	ii, ii
Blood groups:	A, A	O, O
Cross 6	**Group B**	**Group O**
Gametes:	I^B, i	i, i
Children's genotypes:	I^Bi, I^Bi	ii, ii
Blood groups:	B, B	O, O

   Note: Answers can be arranged differently depending on which gamete circle each allele symbol is placed. There are many more crossover combinations possible.
3. (a)

Parent genotype:	I^Ai	ii
Gametes:	I^A, i	i, i
Children's genotypes:	I^Ai, I^Ai	ii, ii
Blood groups:	A, A	O, O

   (b) 50% (0.5) (c) 50% (0.5) (d) 0% (0)
4. (a) Child possible genotypes: I^BI^B or I^Bi
   (b) Woman's possible genotypes: I^AI^A or I^Ai
   Man's genotype: ii
   (c) Therefore the only possible offspring from this couple would have been children with group A or group O. The man making the claim could not have been the father of the child.
5. (a) A or B (b) AB, A, B or O

ISBN: 978-1-927173-94-7

93. Sex Linkage (page 123)

1. Parental genotypes:
   Maternal X^HX^h Maternal gametes X^H, X^h
   Paternal X^HY Paternal gametes X^H, Y
   Possible fertilizations: X^HX^H, X^HX^h, X^HY, X^hY

	Genotypes	Phenotypes
Male children	X^HY	Normal
	X^hY	Hemophiliac
Female children	X^HX^h	Carrier
	X^HX^H	Normal

2. (a) Possible genotypes for the mother: X^HX^H, X^HX^h
   Possible genotypes for the father: X^HY
   Probability that she is a crrier 25% or 1/4
   (b) If the mother is has a normal genotype (X^HX^H) the chance of the offspring being affected is 0%. If the mother is a carrier (X^HX^h) the chance of the offspring being affected is 50%. Because the genotype of the mother is unknown, the overall probability of the offspring being affected is 25%.

3. (a) Genotype X^BX^b
   (b) Phenotype Carrier (normal)
   (c) X^bX^b

4.

Parent:	Normal wife	Affected husband
Parent genotype:	XX	X^RY
Gametes:	X, X	X^R, Y
Children's genotypes:	X^RX, XY	X^RX, XY
Phenotypes:	Affected girl, Normal boy	Affected girl, Normal boy

   (a) Probability of having affected children = 50% or 0.5
   (b) Probability of having an affected girl = 50% or 0.5
   However, all girls born will be affected = 100%
   (c) Probability of having an affected boy = 0% or none

5.

Parent:	Affected wife	Normal husband
Parent genotype:	X^RX	XY
Gametes:	X^R, X	X, Y
Children's genotype:	X^RX, X^RY XX, XY	
Phenotypes:	Affected girl, Affected boy	Normal girl, Normal boy

   Note: Because the wife had a normal father, she must be heterozygous since her father was able to donate only an X-chromosome with the normal condition.
   (a) Probability of having affected children = 50% or 0.5
   (b) Probability of having an affected girl = 25% or 0.25
   However, half of all girls born may be affected.
   (c) Probability of having an affected boy = 25% or 0.25
   However, half of all boys born may be affected.

   Background information for question 6: Sex linkage refers to the location of genes on one or other of the sex chromosomes (usually the X, but a few are carried on the Y). Such genes produce an inheritance pattern which is different from that shown by autosomes:
   - Reciprocal crosses produce different results (unlike autosomal genes that produce the same results).
   - Males carry only one allele of each gene.
   - Dominance operates in females only.
   - A 'cross-cross' inheritance pattern is produced: father to daughter to grandson, etc.

6. Sex linkage (in humans, this usually means X-linkage) is involved in a number of genetic disorders. X-linked disorders are commonly seen only in males, because they have only one locus for the gene and must express the trait. If the sex linked trait is due to a recessive allele, females will express the phenotype only when homozygous recessive. It is possible for females to inherit a double dose of the recessive allele (e.g. a colour blind daughter can be born to a colour blind father and mother who is a carrier), but this is much less likely than in males because sex linked traits are relatively uncommon. Over a hundred X-linked genes are known, including those that control:
   - Fragile X syndrome: The most single-gene cause of autism associated with the expansion of the CGG trinucleotide repeat affecting the FMR1 gene on the X chromosome. About 20% of women are carriers.
   - Antidiuretic hormone production: A version of this gene causes some forms of diabetes insipidus.
   - Muscle development: A rare recessive allele causes Duchene muscular dystrophy.

94. Inheritance Patterns (page 125)

1. Autosomal recessive:
   (a) Punnett square:
   Male parent phenotype:
   Normal, carrier
   Female parent phenotype:
   Normal, carrier

	P	*p*
P	*PP*	*Pp*
p	*Pp*	*pp*

   (b) Phenotype ratio:
   Normal 3 Albino 1

2. Autosomal dominant:
   (a) Punnett square:
   Male parent phenotype:
   Woolly hair
   Female parent phenotype:
   Woolly hair

	W	*w*
W	*WW*	*Ww*
w	*Ww*	*ww*

   (b) Phenotype ratio:
   Normal 1 Woolly 3

3. Sex linked recessive:
   (a) Punnett square:
   Male parent phenotype:
   Normal
   Female parent phenotype:
   Normal, carrier

	X	X_h
X	*XX*	XX_h
Y	*XY*	X_hY

   (b) Phenotype ratio:
   Females:
   Normal 2 Hemophiliac 0
   Males:
   Normal 1 Hemophiliac 1

4. Sex linked dominant:
   (a) Punnett square:
   Male parent phenotype:
   Affected (with rickets)
   Female parent phenotype:
   Affected (with rickets)

	X_R	*X*
X_R	X_RX_R	X_RX
Y	X_RY	*XY*

   (b) Phenotype ratio:
   Females:
   Normal 0 Rickets 2
   Males:
   Normal 1 Rickets 1

95. Problems Involving Monohybrid Inheritance (page 126)

1. 1/2 Ww and 1/2 ww.
   Ratio: 1 wire-haired : 1 smooth haired
   Working: Parental genotypes are Ww X Ww. The test cross of the F1 (to the homozygous recessive by definition) is to a smooth haired dog (ww).
   1/4 of the F1 will be wire-haired (WW). When crossed with ww the result will be all wire-haired dogs (Ww).
   1/2 the F1 will be wire-haired (Ww). When crossed with ww, the result is 1/2 wire-haired and 1/2 smooth-haired.
   1/4 of the F1 will be smooth-haired (ww). When crossed with ww, all offspring will be smooth-haired (ww). Across all progeny, half will be Ww and half will be ww.
2. Probability of black offspring: (2/3 x 1/4=) 1/6 or 0.16
   Working: The parents genotypes are Bb X Bb, and 1/3 of the white offspring (BB) crossed with Bb will result in no black lambs while 2/3 of the white offspring (Bb) crossed with Bb will result in 1/4 black lambs.

ISBN: 978-1-927173-94-7

3. (a) They have an albino child (aa) as well as unaffected ones (AA or Aa), so the parents must both be Aa. Note: There is a 25% chance that any child of theirs will be albino.
(b) The family are all aa.
(c) The albino father must be aa. The mother must be Aa. The three unaffected children are Aa.
Note: There is a 50% chance that any child of theirs will be albino. The observed 3:1 ratio is not surprising, given the small number of offspring.

4. – **Couple #1** genotypes must be X^HX– and X^HY because neither is affected. Their son is affected X^hY. If the mother is X^HX^H they could not have an affected son. If she is X^HX^h, there is a 50% chance that her son will be X^hY.
– **Couple #2** genotypes must be X^HX– and X^hY and their son is X^HY. The father did not pass an X chromosome to his son, so his genotype is irrelevant. If the mother is X^HX^H, all of her sons will be X^HY, but if she is a carrier X^HX^h, there is a 50% chance that her son will be X^hY.
– Either the hospital or the parents could be correct. The answer depends on the genotype of the mothers.

5. There is a possibility that the male is the father of the child as blood group O can result from crossing AO and BO genotypes. However there are also many other possible outcomes. Without more precise testing or knowing the actual genotypes of the male and female it is impossible to conclusively say the male is the father of the child.

96. Pedigree Analysis (page 127)

1. Pedigree chart of your family: The chart should be drawn up using the appropriate symbols.
2. Autosomal recessive traits
(a) Genotypes of individuals on the chart:

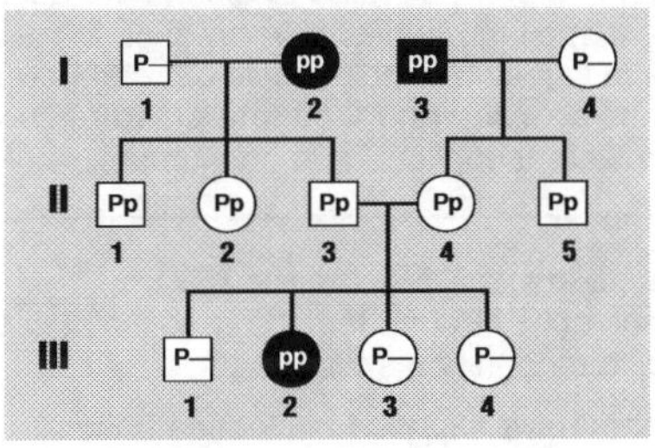

(b) The genotypes of parents II-3 and II-4 have to be carriers (heterozygous) because they produced an affected offspring (homozygous recessive). Alternatively, we know that they must be carriers because each had an affected parent (I-2 and I-3).

3. Sex-linked recessive traits
(a) Genotypes of individuals on the chart:

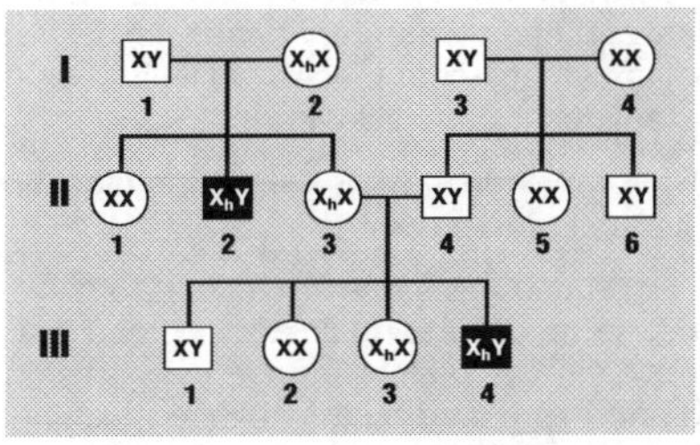

(b) Males have only one X chromosome. If that single chromosome carries the affected allele, then it will be expressed. Males cannot be heterozygous, since they can only carry one copy of the gene.

4. Autosomal dominant traits
(a) Genotypes of individuals on the chart:
Teacher's note: Parent I-2 has to be heterozygous (Ww) since some of the offspring are normal (ww). This could not occur if I-2 was homozygous dominant (WW).

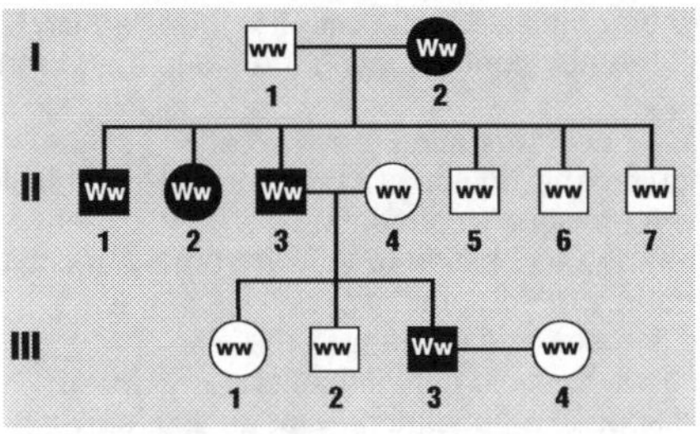

(b) Each affected individual has an affected parent.

5. Sex-linked dominant traits
(a) Genotypes of individuals on the chart:

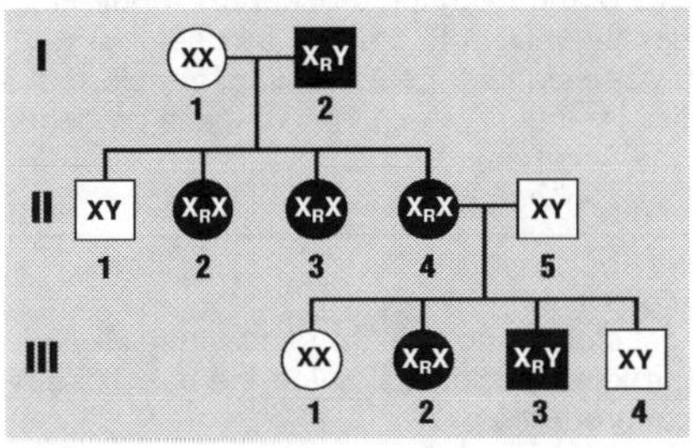

(b) Females have two X chromosomes, and therefore have a greater probability that one of them will carry the affected gene.

97. Amazing Organisms, Amazing Enzymes (page 129)

1. PCR only works as a viable process when the enzymes being used do not have to be replaced with every cycle. Up until the mid 1980s no enzymes were available that could do this. It was only after Taq polymerase was isolated that PCR became a common place technique.
2. Taq polymerase is an DNA polymerase enzyme that functions at high temperatures. This means it can remain in the PCR mix during the heating cycles and can be used again during each cycle of DNA replication. Without it, polymerase enzymes must be added after each cycle of heating.
3. Studying the lifestyles of diverse organisms enables us to understand how they cope with environmental challenges. This leads to investigation and isolation of unique and potentially useful molecules, which can be used in different situations, e.g. in industrial processing, in food processing, or in medicine.

98. Polymerase Chain Reaction (page 130)

1. PCR produces large quantities of 'cloned' DNA from very small samples. Large quantities are needed for effective analysis. Minute quantities are often unusable.
2. A double stranded DNA is heated to 98°C for 5 min, causing the two strands to separate. Primer starting DNA polymerase are added to the sample. This is then incubated at 60°C for a few minutes during which time, complementary strands are created using each strand of the DNA sample as a template. The process is repeated about 25 times, each time the number of templates doubles over the previous cycle.
3. (a) Forensic samples taken at the scene of a crime (for example, hair, blood, or semen).
(b) Archeological samples from early human remains.
(c) Samples taken from the remains of prehistoric organisms preserved in ice, mummified, preserved in amber, tar pits etc.
4. Do this on a calculator by pressing the 1 button (original sample) and then multiplying by 2 repeatedly (to simulate each cycle). Most calculators will not display more than about 8 digits. Alternatively, calculate using a spreadsheet or use the exponent function the calculator (2^{10} and 2^{25}).
(a) 1024 (b) 33,554,432 (33.5×10^6)

ISBN: 978-1-927173-94-7

5. (a) It would be amplified along with the intended DNA sample, thereby contaminating the sample and rendering it unusable.
   (b) Sources of contamination (any two of):
   Dirty equipment (equipment that has DNA molecules left on it from previous treatments).
   DNA from the technician (dandruff from the technician is a major source of contamination!)
   Spores, viruses and bacteria in the air.
   (c) Precautions to avoid contamination (any two of):
   Using disposable equipment (pipette tips, gloves).
   Wearing a head cover (disposable cap).
   Use of sterile procedures.
   Use of plastic disposable tubes with caps that seal the contents from air contamination.
6. (a) and (b) any of the following procedures require a certain minimum quantity of DNA in order to be useful: DNA sequencing, gene cloning, DNA profiling, transformation, making artificial genes. Descriptions of these procedures are provided in the workbook.

---

99. Gel Electrophoresis (page 132)

1. Purpose of gel electrophoresis: To separate mixtures of molecules (proteins, nucleic acids) on the basis of size and other physical properties.
2. (a) The frictional (retarding) force of each fragment's size (larger fragments travel more slowly than smaller ones).
   (b) The strength of the electric field (movement is more rapid in a stronger field).

   Note: The temperature and the ionic strength of the buffer can be varied to optimize separation.
3. The gel is full of pores (holes) through which the fragments must pass. Smaller fragments will pass through these pores more easily (with less resistance and therefore faster) than larger ones.

---

100. DNA Profiling Using PCR (page 133)

1. STRs (microsatellites) are non-coding nucleotide sequences (2-6 base pairs long) that repeat themselves many times over (repeats of up to 100X). The human genome has numerous different STRs; equivalent sequences in different people vary considerably in the numbers of the repeating unit. This property can be used to identify the natural variation found in every person's DNA since every person will have a different combination of STRs of different repeat length.
2. (a) Gel electrophoresis: Used to separate the DNA fragments (STRs) according to size to create the fingerprint or profile.
   (b) PCR: Used to make many copies of the STRs. Only the STR sites are amplified by PCR, because the primers used to initiate the PCR are very specific.
3. (a) Extract the DNA from sample. Treat the tissue with chemicals and enzymes to extract the DNA, which is then separated and purified.
   (b) Amplify the microsatellite using PCR. Primers are used to make large quantities of the STR.
   (c) Run the fragments through a gel to separate them. The resulting pattern represents the STR sizes for that individual (different from that of other people).
4. To ensure that the number of STR sites, when compared, will produce a profile that is effectively unique (different from just about every other individual). It provides a high degree of statistical confidence when a match occurs.

---

101. Forensic Applications of DNA Profiling (page 135)

1. Profiles of everyone involved must be completed to compare their DNA to any DNA found at the scene and therefore eliminate (or implicate) them as suspects.
2. The alleged offender is not guilty. The alleged offender's DNA profile does not appear in the DNA collected at the crime scene nor does it appear in the DNA database. Profile E's DNA is found at the scene.
3. CSF1PO occurs in only about 3% of the population (0.0294 out of 1). Only 3 out of every 100 people will have this allele. Allele 12 is found in more than 34% of the population. A match between allele 9 is therefore ten times less likely than a match between allele 12, and unlikely to happen unless the father is the parent of the child.

---

102. What is Genetic Modification? (page 136)

1. Organisms may be genetically modified through:
   - The addition of a foreign gene, e.g. human insulin gene inserted into bacteria or yeast for the commercial production of human insulin.
   - Alteration of an existing gene so that a protein is expressed at a higher rate or in a different way. This GM technique is used in gene therapy.
   - Deletion or inactivation of an existing gene, e.g. the Flavr-Savr tomato which has had its ripening gene switched off. Gene inactivation also produces 'knock-out mice' which are used to study the physiological effects of particular genes.
2. (a) Gene therapy: A need/desire to find cures/treatments for genetic diseases (e.g. cystic fibrosis) especially when they can be attributed to a single gene mutation and so are more easily targeted.
   (b) Transgenic organisms: A need to accelerate traditional breeding programmes by direct manipulation of the genome and a desire to improve the usefulness of livestock and crops by increasing production and reducing susceptibility to diseases and pests. There has also been a desire to produce new protein products by providing novel genes to make plants and animals into biofactories.
   (c) The extension of shelf life: A need to extend storage times to prevent spoilage and enable produce to reach its destination in saleable condition without deterioration. This means less spoilage and therefore greater economic benefits for growers, suppliers, and consumers.

---

103. Making Recombinant DNA (page 137)

1. Restriction enzymes cut DNA into lengths or to isolate genes by cutting at specific recognition sites.
2. Sticky end fragments have exposed nucleotide bases at each end, whereas blunt end fragments have no exposed nucleotide bases. Sticky ends are joined via the complementary base pairing of the exposed bases.
3. Having many different kinds of restriction enzymes allows DNA to be cut at many different recognition sites and so produce a variety of sticky or blunt ends. This allows for a better ability to isolate and join different regions of the DNA.
4. (a) The two single-stranded DNA molecules are recombined into a double-stranded molecule. This is achieved by H-bonding between complementary bases.
   (b) DNA ligase joins two adjacent pieces of DNA by linking nucleotides in the sticky ends.
5. It joins together DNA molecules, while restriction digestion (by restriction enzymes) cuts them up.
6. All organisms on Earth use the same DNA code to store information and the same cellular machinery to read the information and express it. Any DNA from any organism can therefore be read and expressed by the cellular machinery of any other organism into which the DNA is spliced.

---

104. Applications of GMOs (page 139)

1. Student's own research activity based on information provided in the student workbook or in other texts.

---

105. *In Vivo* Gene Cloning (page 140)

1. *In vivo* methods make it simpler to produce the protein product because the recipient of the gene (the bacterium) will express the gene as its protein product and this product can

ISBN: 978-1-927173-94-7

then be harvested.

2. One would not use bacteria to clone a gene if the purpose of the gene cloning was simply to amplify the DNA, e.g. for forensic or diagnostic purposes, and protein expression was not required. PCR alone is a simple automated process.

3. The gene is prepared by removal of introns. At the same time, an appropriate vector (e.g. plasmid) is isolated. Both the gene and plasmid are treated with the same restriction enzyme to produce identical sticky ends. The DNA fragments are mixed in the presence of DNA ligase and anneal (DNA ligation).

4. 2 x 24 = 48 replications in twenty four hours. Number of bacteria and therefore plasmids = 248 or 2 multiplied by itself 48 times = 281,474,976,710,656 copies.

5. (a) On the plasmid
   (b) The bacteria containing the plasmids with the human gene will die as they don't have tetracycline resistance.
   (c) The dead colonies can be traced back to the original colony they were sampled from. These colonies have the human gene.

6. A *gfp* marker is preferable over antibiotic resistance genes because antibiotic resistance genes encourage the undesirable spread of antibiotic resistance through bacterial populations. Apart from this, a gfp marker system is much simpler, as the marked colonies can be quickly identified by fluorescence under UV light.

7. (a) Bacteriophages can insert their DNA into the bacteria that they infect (the host species).
   (b) Bacteriophages can be used to clone a gene by firstly inserting the gene into the bacteriophage. The bacteriophage is then introduced into a suspension of bacteria, which it infects. The bacteria then begin making copies of the virus and therefore clone the target gene.

106. Using Recombinant Bacteria (page 142)

1. Chymosin (an enzyme) is used to coagulate milk into curds in the production of cheese.

2. Traditional source of chymosin was from the stomachs of young (suckling) calves.

3. (a) The amino acid sequence of a gene protein product and its mRNA sequence can be determined.
   (b) Reverse transcriptase can synthesize DNA from mRNA to construct a protein-coding gene.
   (c) Restriction enzymes can be used to cut DNA at specific sites and joined again with DNA ligase. Genes of interest can be isolated and inserted into a vector, e.g. plasmid).
   (d) Bacteria can take up the modified plasmid and replicate it in culture.
   (e) Large quantities of the protein can be produced by growing the bacteria in large batches.

4. Chymosin gene was isolated by first determining its amino acid sequence and from that, the mRNA coding sequence. Once this is known, a gene probe can be constructed and used to locate the mRNA of interest. Reverse transcriptase is used to produce DNA strand, which can then be amplified using PCR. This technique can be applied to isolating any gene of interest, once its protein product is known.

5. Advantages of using GE chymosin, (a)-(c) any 3 of:
   - The chymosin produced is identical to chymosin from its natural source.
   - The chymosin can be produced without the unnecessary slaughter of calves (a welfare issue for many people) and is suitable for vegetarians.
   - The chymosin is 80-90% active ingredient and is thereby significantly purer as natural rennet, which contains only between 4-8% active chymosin.
   - The chymosin can be produced on demand, in the quantities required. This makes it cost effective.

6. Fungi are eukaryotes. The cells are larger and their secretory pathways are more similar to those of humans than those of bacteria such as *E.coli*.

107. Golden Rice (page 144)

1. The genes for two different enzymes involved in beta carotene synthesis are taken from two different sources and inserted into the nuclear genome of a rice plant. Expression of the gene under the control of an endosperm specific promoter results in production of beta carotene in the edible portion of the rice plant.

2. The expression of the genes is controlled by a promoter specific to the endosperm, so the genes will only be expressed in that tissue.

3. *Agrobacterium tumefaciens* is a natural plant pathogen and can transfer genes as a consequence of infecting a host plant. The tumour-inducing *Ti* plasmid can be modified to delete the tumour-forming gene and insert a gene for a desirable trait.

4. (a) Production and consumption of beta-carotene rich rice could alleviate or prevent diseases related to vitamin A deficiency (e.g. night blindness and susceptibility to infection as a result of low immunity). Beta carotene is a precursor to vitamin A.
   (b) Improved nutrition through GM rice will be viable only if the diet in targeted regions is also adequate with respect to fat intake. In some impoverished regions this will not be the case, as diet is inadequate across a wide range of food groups, including fat and protein.

5. (a) More grains per head, larger grains.
   (b) Faster maturation time, even in marginal conditions.
   (c) Production of natural toxins, thicker seed cat, low palatability to insects, shift in timing of vulnerable stage.

108. Production of Insulin (page 146)

1. (a) High cost (extraction from tissue is expensive).
   (b) Non-human insulin is different enough from human insulin to cause side effects.
   (c) The extraction methods did not produce pure insulin so the insulin was often contaminated.

2. Mass production of human proteins using *E. coli* facilitates a low cost, reliable supply for consumer use. The insulin protein is free of contaminants and, because it is a human protein, the side effects of its use are minimized.

3. The insulin is synthesized as two (A and B) nucleotide sequences (corresponding to the two polypeptide chains) because a single sequence is too large to be inserted into the bacterial plasmid. Two shorter sequences are small enough to be inserted (separately) into bacterial plasmids.

4. The β-galactosidase gene in *E.coli* carries a promoter for transcription, so the synthetic genes must be tied to that gene in order to be transcribed.

5. (a) Insertion of the gene: The yeast plasmid is larger and can accommodate the entire synthetic nucleotide sequence for the A and B chains as one uninterrupted sequence.
   (b) Secretion and purification: Yeast, a eukaryote, has secretory pathways that are more similar to humans than those of a prokaryote and β-galactosidase is not required for gene expression. Secretion of the precursor insulin molecules is therefore less problematic. Purification is simplified because removal of β-galactosidase is not required and the separate protein chains do not need to be joined.

109. Food for the Masses (page 148)

1. The bacterium because it uses only one enzyme to facilitate multiple reactions. It is therefore simpler to use in the production of the modified plant.

2. The gene can be isolated by first identifying the associated enzyme and its amino acid sequence. From this, the mRNA sequence can be identified. The correct mRNA molecules can then be extracted and reverse transcriptase used to copy the mRNA into DNA. This can then be amplified.

   Or - Once the DNA sequence is identified, PCR primers can be produced that will anneal to the start sequences of the gene. The PCR product will be the targeted gene.

ISBN: 978-1-927173-94-7

Or - The gene can be identified from its protein product (as above). This can then be cut from the chromosome using restriction enzymes and amplified.

3. *Agrobacterium*. This bacterium can transfer DNA into plants, so that the cells will contain recombinant DNA.
4. A plasmid (the *Ti* plasmid) is removed from the *Agrobacterium*. Using restriction enzymes the plasmid is cut and the target DNA inserted (the tumor forming gene is removed). DNA ligase is used to attach the target DNA to the plasmid. The plasmid is then replaced into the *Agrobacterium*.
5. (a) *Agrobacterium* transfers the recombinant plasmid to the cells of the target plant.
   (b) The best stage of development for transformation is while the plant is in embryonic form. In this way a larger proportion of the plant cells will be affected and take up the new DNA. This will cause a much better result in the adult plant.
6. Transformed plants can be identified if an extra gene is inserted along with the target gene. This is normally a gene for drug resistance. Plants grown on agar impregnated with the chemical will grow only if they have taken up the new DNA. Those without won't grow.
7. A large number of plants can be produced using plant tissue culture or vegetative propagation. In this way many plants can be produced, which will lead to rapid dissemination of the transgenic stock.

110. The Ethics of GMO Technology (page 150)

1. (a)-(c) any of (not an exclusive list as there are other benefits):
   - Herbicides could be used to control weeds without harming the crop, increasing productivity.
   - Crops can be grown outside of their natural tolerance range, e.g tropical crops in temperate climates, increasing food production globally.
   - Crops can be intrinsically pesticide resistant so pesticide use can be reduced.
   - Storage time in harvested crops can be improved, reducing waste and improving economic gain for growers.
   - GMOs can be used as biofactories to produce a cost effective safe supply of high value product, e.g. medicines.
2. (a)-(c) any of (not an exclusive list as there are other risks):
   - Herbicide resistance may be spread to weeds by viral or bacterial vectors that infect plants. Resistance encourages overuse of herbicide chemicals.
   - Modified plants in a new environment may become pest species. Developing economies that rely on tropical cash crops may suffer as a result of competition from economically strong countries.
   - A greater ability to grow crops in marginal areas may result in previously uncultivated land (e.g. wetlands or forest) being used for agriculture, thus reducing natural areas and harming biodiversity.
   - The use of transgenic livestock as biofactories carries a risk of cross-species transfer of disease and is associated with potential welfare issues for the animal expressing the foreign protein.
3. Students generate their own answers. Students could discuss where in the production pathway blame should reasonably be attributed or how production can be maximized without harming the environment.
4. Note: This question was not intended to imply that ethical or moral concerns are any less real or valid than biological ones. It was merely an exercise in identifying the nature of the biological concerns.

   (a) Some points for discussion are:
   - The GM product or the GMO could have an undesirable effect on humans or other organisms.
   - That the genetic modification would spread uncontrollably into other organisms (breeding populations of the same or different species).
   - Consumer choice is denied unless adequate labelling protocols are in place. If everything contains GM products, there is no consumer choice.
   - General fear of what is not understood (fear of real or imagined consequences).
   - Objections on the grounds that it is ethically and morally wrong to tamper with the genetic make-up of an organism.
   - Generation of monopolies where large companies control the rights to seed supplies & breeding stock.

   (b) Those that pose a real biological threat are:
   - Amongst GM plants, the indiscriminate spread of foreign genes.
   - Unusual physiological reactions, e.g. allergies, to novel proteins.
   - Animal welfare issues may be justified if genetic modification causes impaired health or quality of life.

111. Natural Clones (page 152)

1. (a) The clones are genetically identical
   (b) The population will adapt to change slowly (through accumulated mutations) as there is little capacity for variation in a population of genetically identical individuals.

112. Cloning by Embryo Splitting (page 153)

1. Making an exact copy of a cell or animal by replicating it from a single parent. This is achieved either by splitting embryos at an early stage of embryonic development (embryo splitting), or by nuclear transfer techniques.
2. Embryo splitting allows breeders to produce multiple embryos that can be implanted into multiple surrogate mothers, producing multiple offspring.
3. Cloning high milk yielding cows will reduce the time it takes to produce a high producing herd. Multiple offspring will be produced quickly from the most valuable phenotypes. The genetic gain will be more rapid than that achieved by traditional selective breeding methods.
4. Repeated exclusive use of embryo splitting will reduce the total pool of genetic diversity from which to select new breeds/ strains/varieties.

113. Cloning by Somatic Cell Nuclear Transfer (page 154)

1. Adult cloning is the production of genetically identical individuals from the non-embryonic (somatic) tissue of a known phenotype.
2. (a) Switching off genes in the donor cell:
   Induced by low nutrient medium.
   (b) Fusion of donor and enucleated egg:
   Induced by a short electric pulse.
   (c) Activation of the cloned cell:
   Induced by a second gentle electric pulse or by chemical means. A time delay of about 6 hours improves the success of the egg activation process, probably through the prolonged contact of the chromatin with (unknown) cytoplasmic factors.
3. Potential applications (a) and (b) any two of:
   - Production of clones from a proven phenotype that can quickly be disseminated to commercial herds.
   - Rapid production of transgenic animals that produce a particular product (e.g. a pharmaceutical secreted in milk), in order to respond to a market demand.
   - Conservation of rare breeds. It is hoped that cloning will be integrated into zoo management programmes. By retaining the tissues of individuals before they die, some of the genetic diversity of rare species can be retained. It may even be possible to restore species that are on the verge of extinction using cloning technology.

114. Chapter Review (page 156)

No model answer. Summary is the student's own.

115. KEY TERMS: Did You Get it? (page 158)

1. Alleles (E), autosome (J), diploid (Q), DNA profiling (M),

ISBN: 978-1-927173-94-7

dominant (I), gel electrophoresis (A), genetic modification (D), genotype (K), karyotype (L), meiosis (F), monohybrid cross (C), mutation (G), non-disjunction (O), PCR (P), phenotype (N), Punnett square (H), recessive (B)

2. Mutations are the ultimate... ...source of new alleles.
Alleles are variations... ...of a gene.
A person carrying two of the same alleles (one on each homologous chromosome)... ...is said to be homozygous
If the person carries two different alleles... ...for the gene, they are heterozygous.
Alleles may be... ...dominant or recessive
A dominant allele... ...always expresses its trait whether it is homozygous or heterozygous
A recessive allele only... ...expresses its trait if it is homozygous.

116. Components of an Ecosystem (page 160)

1. A community is a naturally occurring group of organisms living together as an ecological entity. The community is the biological part of the ecosystem. The ecosystem includes all of the organisms (the community) and their physical environment.
2. The biotic factors are the influences that result from the activities of living organisms in the community whereas the abiotic (physical) factors comprise the non-living part of the community, e.g. climate.
3. (a) Population (c) The community
(b) Ecosystem (d) Physical factor
4. (a) A group of organisms that can potentially interbreed to produce fertile offspring.
(b) The populations have not been separated for long enough to become reproductively isolated. If they came into contact they could interbreed to produce fertile offspring.
5. (a) Random sampling is the sampling of a population so that every possible sample of a given size has the same chance of selection.
(b) Random sampling is recommended because it avoids bias in the results. As long as enough samples are taken, random sampling gives an unbiased, representative picture of the whole population.

117. The Stability of Ecosystems (page 162)

1. Adding algal nutrients to the growth chambers increased the population growth of the copepod for the first 20 days, presumably because they responded to increased growth of their algal food. 2% and 10% algal growth chambers produced stable populations after 20 days, although 2% did begin to decline slightly from then on. The population in the 20% growth chamber grew rapidly but declined quickly after 20 days. It never reached a stable state.
2. – Assumption that the algae in the chambers were acceptable food for the copepods.
– There is the assumption that the algal growth is directly related to the growth nutrients.
– An assumption that male and female copepods were added or females with eggs were added.

118. Quadrat Sampling (page 163)

1. Mean number of centipedes captured per quadrat:
Total number centipedes ÷ total number quadrats
= 30 individuals ÷ 37 quadrats
= 0.811 centipedes per quadrat
2. Number per quadrat ÷ area of each quadrat
0.811 ÷ 0.08 = 10.1 centipedes per m^2
3. Clumped or random distribution.
4. Presence of suitable microhabitats for cover (e.g. logs, stones, leaf litter) may be scattered.

119. Sampling a Rocky Shore Community (page 164)

1. The results per se are not particularly important, but it is important to understand the method and its limitations. The results will vary depending on a group's agreed criteria for inclusion of organisms in a given quadrat (e.g. when and how an organism is counted when it is partly inside a quadrat). Note: Some algae are almost obscured by organisms or have other algae on top of them.
6. Typical results (total for each category) are:

	A	B	C	D	Direct count
Barnacle:	9	6	6	16	80
Oyster borer:	0	0	1	1	3
Chiton:	1	0	1	0	3
Limpet:	0	3	0	0	6
Algae:	27	18	15	13	101

7. Typical results for calculated population density based on A-D and on a direct count (question 8b):

	A	B	C	D	Direct count
Barnacle:	1667	1111	1111	2963	2469
Oyster borer:	0	0	185	185	93
Chiton:	185	0	185	0	93
Limpet:	0	556	0	0	185
Algae:	5000	3333	2778	2407	3117

NB: Area of 6 quadrats = (0.03 x 0.03) x 6 = 0.0054 m^2
Area of total sample area = 0.18 x 0.18 = 0.0324 m^2

8. (a) Once the quadrats have been laid, the animals moving from one quadrat to another could be counted twice. The quadrat could involve the placement of physical barriers between each quadrat (what about the invertebrates directly underneath?). Possibility of exposing the entire area and photographing it for later analysis.
(b) Densities calculated on direct count in the last column of the table above. Students should be aware of the dangers of extrapolating data from a small sample. Including or excluding single individuals can have a large effect on the density calculated, particularly where species are present at low densities. Extension: Groups could combine data to see if they get a more representative sample (i.e. closer to the direct count).

121. Chi-Squared Exercise in Ecology (page 167)

1. There is no association between lesser pond sedge and marsh bedstraw.
2. LPS and MBS were found together in 11% of quadrat samples. When LPS was counted present, MBS was present less often than it was absent. When LPS was counted absent MBS was absent more often than it was present.
3.

	LPS present (1)	LPS absent (0)	Total
MBS present (1)	5.88	8.12	14
MBS absent (0)	36.12	49.88	86
Total	42	58	100

4.

Category	O	E	O–E	$(O-E)^2$	$\frac{(O-E)^2}{E}$
LPS 1/MBS 1	11	5.88	5.12	26.2	4.45
LPS 0/MBS 1	3	8.12	-5.12	26.2	3.2
LPS 1/MBS 0	31	36.12	-5.12	26.2	0.73
LPS 0/MBS 0	55	49.88	5.12	26.2	0.53
					$\Sigma = 8.91$

ISBN: 978-1-927173-94-7

5. df = 1
6. 0.005
7. The null hypothesis of no association can be rejected (at the p = 0.005 level).
8. LPS and MBS are found associated more often than you would expect by chance alone.

122. Food Chains (page 168)

1. (a) The sun.
   (b) Refer to the diagram, below

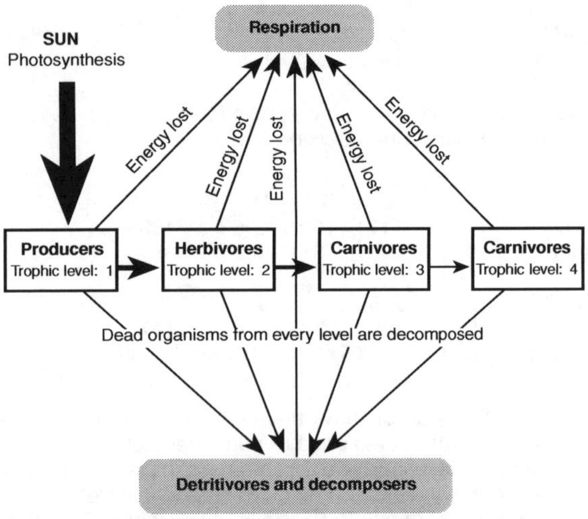

Some secondary consumers feed directly off decomposer organisms

2. (a) Producers: obtain energy directly from the sun via photosynthesis.
   (b) Consumers: obtain energy by eating other organisms.
   (c) Detritivores: obtain energy from eating dead organic matter.
   (d) Saprotrophs: obtain energy by extracellular digestion of dead material.

123. Food Webs (page 169)

1. Energy passes through ecosystems by being transferred from producers to consumers. Some energy is lost at each transition in metabolic wastes and as heat.
2. (a) The amount of energy available to each successive trophic level decreases along the food chain.
   (b) Energy is lost as heat and as organic waste at each trophic level. This energy is not available to the next trophic level. Heat energy cannot be used by organisms although the energy in organic waste is available to detritivores and decomposers.
3. In a grazing food web, energy moves from producers to primary consumers and then to secondary consumers. This chain of energy transfer can continue several times, but eventually ends. All consumer groups provide energy to decomposer levels. In a detrital food web, producers provide energy as dead plant material, and the primary consumers are decomposers. Energy flows back and forth between decomposers and detritivores but herbivores and carnivores do not feature.

124. Constructing a Food Web (page 170)

1. Some food chain examples as below (there are others)
   (a) Algae → zooplankton → diving beetle
   (b) Algae → zooplankton → stickleback → pike
   (c) Macrophyte → great pond snail → herbivorous water beetle → stickleback → pike
   (d) Macrophyte → carp → pike
   (e) Algae → mosquito larva → Hydra → dragonfly larva → carp → pike
   (f) Macrophyte → herbivorous water beetle → carp → pike
   (g) Algae → zooplankton → Asplanchna → leech → dragonfly larva → carp → pike
   (h) Detritus → Paramecium → Asplanchna → leech → dragonfly larva → carp → pike
   (i) Detritus → great pond snail → leech → dragonfly larva → carp → pike
   (j) Detritus → Paramecium → mosquito larva → Hydra → dragonfly larva → carp→ pike
2.

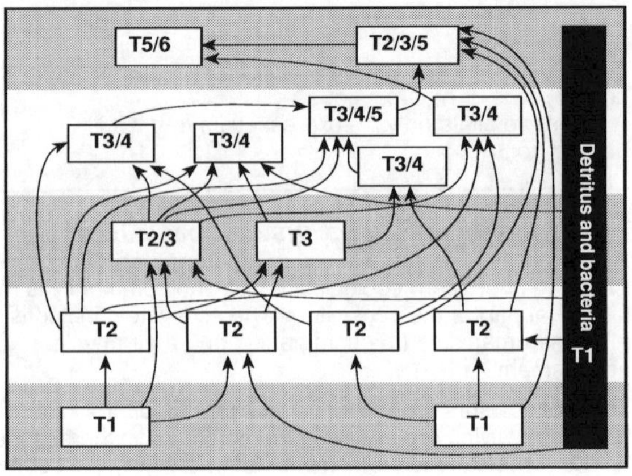

125. Energy Flow in an Ecosystem (page 172)

1. (a) 14,000 (b) 180
   (c) 35 (d) 100
2. A. Photosynthesis
   B. Eating/feeding/ingestion
   C. Respiration
   D. Export (lost from this ecosystem to another)
   E. Decomposers and detritivores feeding on other decomposers and detritivores
   F. Radiation of heat to the atmosphere
   G. Excretion/egestion/death
3. (a) 1,700,000 ÷ 7,000,000 x 100 = 24.28%
   (b) It is reflected. Plants appear green because those wavelengths are not absorbed. Reflected light falls on other objects as well as back into space.
4. (a) 87,400 ÷ 1,700,000 x 100 = 5.14%
   (b) 1,700,000 - 87,400 = 1,612,600 (94.86%)
   (c) Most of the energy absorbed by producers is not used in photosynthesis. This excess energy, which is not fixed, is lost as heat (although the heat loss component before the producer level is not usually shown on energy flow diagrams). Note: Some of the light energy absorbed through accessory pigments such as carotenoids widens the spectrum that can drive photosynthesis. However, much of accessory pigment activity is associated with photoprotection; they absorb and dissipate excess light energy that would otherwise damage chlorophyll.
5. (a) 78,835 kJ
   (b) 78,835 ÷ 1,700,000 x 100 = 4.64%
6. (a) Decomposers and detritivores
   (b) Transport by wind or water to another ecosystem (e.g. blown or carried in water currents).
7. (a) Low oxygen or anaerobic, low temperature, low moisture.
   (b) Energy remains locked up in the detrital material and is not released.
   (c) Geological reservoir:

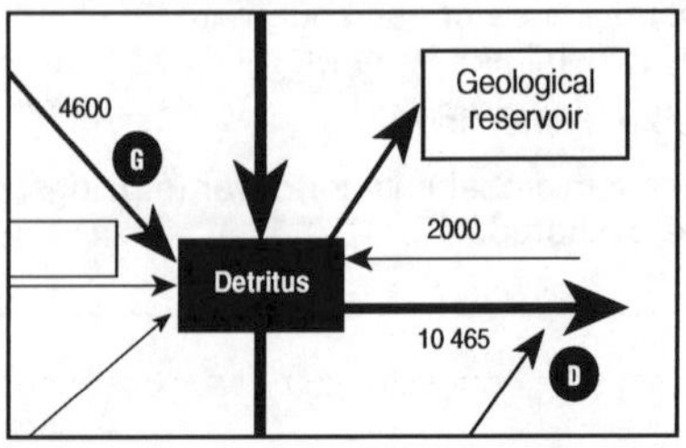

ISBN: 978-1-927173-94-7

(d) Oil (petroleum) and natural gas, formed from the remains of marine plankton. Coal and peat are both of plant origin; peat is partly decomposed, and coal is fossilized.

8. (a) 87,400 → 14,000: 14,000 ÷ 87,400 x 100 = 16%
(b) 14,000 → 1600: 1600 ÷14,000 x 100 = 11.4%
(c) 1600 → 90: 90 ÷ 1600 x 100 = 5.6

126. Ecological Pyramids (page 174)

1. (a) 8690 → 142 = 8548 kJ = 1.6%
(b) 142 → 12 = 130 kJ = 8.5%
(c) Energy passed on from producers to primary consumers is less than the expected 10% because a lot of energy is diverted to the decomposers.
(d) Decomposers
(e) In a plankton community, turnover times (generation times of organisms) are very short and there is a lot of dead material both in the water and on the bottom. This provides a rich energy source to support a large biomass of decomposers.

127. Nutrient Cycles (page 175)

1. Energy can be lost from the ecosystem as heat (organisms cannot use heat to produce useful chemical work). It must therefore be constantly added to the system (as sunlight). The amount of energy available to the system also decreases through trophic levels. Nutrients (being matter) can be constantly cycled through an ecosystem. However unlike energy, the amount of matter (nutrients) in ecosystems is finite and limited.

2. (a) Bacteria are able to make conversions to and from elements and their ionic states. This gives plants and animals access to nutrients that they would otherwise not have (i.e. increases bioavailability).
(b) Fungi decompose organic matter, returning nutrients to the soil where plants and bacteria can access them. They are also able to convert some nutrients into more readily accessible forms.
(c) Plants can make their own food and, when they die, add this to the soil in the form of nutrients that can be broken down and used by bacteria and fungi. They also provide browsing animals with nutrients when they are eaten.
(d) Animals break down materials from plants, fungi and bacteria and return then to the soil with their wastes and when they die, allowing the nutrients incorporated in them to re-enter the cycle.

128. The Carbon Cycle (page 176)

1. Arrows can be added for the points (a)-(d) as follows:
(a) Dissolving of limestone by acid rain: Arrow from the limestone layer to atmospheric CO_2.
(b) Release of carbon from the marine food chain: Arrows (labelled respiration) from marine organisms (dolphins, algae, fish) to atmospheric CO_2.
(c) Mining and burning of coal: Arrow from the coal seam to atmospheric CO_2.
(d) Burning of plant material: Arrow (labelled combustion) from the trees and/or grassland to atmospheric CO_2.

2. (a) Respiration (stepwise oxidation of glucose), combustion (rapid oxidation of organic substances accompanied by heat), and release of methane by methanogenic archaeans.
(b) Respiration and combustion release CO_2. Methanogenic archaeans release methane which is oxidized to CO_2.

3. (a) - (d) in any order: Atmosphere, coal, limestone, and oil and natural gas.

4. (a) Photosynthesis (b) Respiration

5. Carbon would eventually be locked up in the bodies (remains) of dead organisms. Dead matter would not rot. Possible gradual loss of CO_2 from the atmosphere.

6. (a) Coal: Plant material trapped under sediment in swampy conditions millions of years ago.
(b) Oil: Marine plankton rapidly buried in sediment mya.
(c) Limestone (also chalk = fine limestone): Shells of molluscs, skeletons of coral and other marine organisms with skeletons of calcium carbonate piled upon seabeds and compressed.
(d) Peat: Partly decayed vegetation formed because of acidic or anaerobic conditions.

7. (a) Methanogenic archaea decompose organic matter (e.g. peat) in anaerobic conditions and produce methane which is released into the atmosphere.
(b) Cattle belching or flatulence.

8. (a) Humans deplete these fossil fuel reserves through mining (fossil fuels provide readily available energy).
(b) The burning of fossil fuels increases the amount of carbon dioxide in the atmosphere, contributing to the rise in global temperatures. Burning also increases levels of air pollution.
(c) Minimizing fossil fuels use through the use of alternative, environmentally clean sources of energy (solar energy, wind energy). Making sure that when fossil fuels are burnt, that combustion is as clean (complete) as possible, to minimize pollution.

129. The Greenhouse Effect (page 178)

1. A greenhouse gas is a gas in the atmosphere that causes the retention of heat in the atmosphere.

2. The greenhouse effect is the retention of heat from the Sun by the Earth's atmosphere.

3. (a) Carbon dioxide: 41.2% increase
(b) Methane: 156.6% increase
(c) Nitrous oxide: 20.0% increase

4. Water is often left out of greenhouse gas table as the amount of water vapour in the atmosphere is related to temperature and therefore dependent on other greenhouse gases.

130. Global Warming (page 179)

1. Increased levels of carbon dioxide, methane, and nitrous oxides act as additional thermal blankets around the Earth, allowing the sun's energy to reach the Earth's surface, but preventing the heat escaping. This means that the Earth slowly heats up. Note: The atmospheric concentrations of these gases have increased dramatically above pre-industrial levels since 1750. These levels are considerably higher than at any time during the last 650,000 years (the period for which reliable data has been extracted from ice cores) and are correlated with a rise in global temperature and documented sea level rises. While correlation does not mean cause and effect, the majority of climate scientists accept the theory that the increase in anthropogenic greenhouse gas emissions is causing the rise in the Earth's temperature.

2 (a) CO_2 levels increased sharply shortly afterwards.
(b) The invention and mass use of the steam engine to provide mechanical power saw huge amounts of coal being burned to fuel them, raising atmospheric CO_2 levels.

3. Models must be revised to take account of changing climatic conditions, new information (changing inputs), better computing power, and the development of more complex models (incorporating more variables) that more realistically describe and predict climate change scenarios.

4. Students own research and arguments. Students need to take into account the accuracy of sceptical arguments, including the data used and its source.

131. Global Warming and Effects on Biodiversity (page 181)

1. Global warming will result in an increased frequency of weather extremes (floods, droughts etc) and a loss of land as coastal areas are inundated. Erosion rates may also increase as a result. Glacial retreats will reduce water supplies and snow lines will increase in altitude. Climate changes may shift

ISBN: 978-1-927173-94-7

the governing physical environment in certain regions (and consequently cause a shift in predominant vegetation). Ocean pH will also fall as a result of CO_2 absorption (again, with consequent changes in biotic communities).

2. (a) Depending on the crop involved, growing range may shrink, expand, or shift. Crop plants may be affected more by higher night temperatures than by higher daytime temperatures. High night temperatures affect the ability of some crop plants such as rice to set seed and fruit. This will cause a reduction in yield, and so also a decrease in the amount of seed available.
   (b) Farmers may adjust by planting different crops in some areas, e.g. crops that are able to grow and set seed at higher temperatures. New crop strains may be able to be developed for the higher temperature ranges.
3. Evidence suggests that insect populations will be affected by global warming. Plant fossil data shows that browse damage peaked during past thermal maxima, suggesting that insect populations will increase in size as temperatures rise.
4. (a) Migratory birds in the northern hemisphere are now not travelling as far south during the winter months (as higher latitudes become more hospitable) and they are making their migrations north up to two weeks earlier than usual.
   (b) Migratory birds may arrive at feeding grounds before the main food supply is ready. Plants with daylength-dependent flowering may not yet be flowering and, as a consequence, insects (and seeds and fruits) may not be in the abundance required to feed the migrants. In addition, the distribution of food resources may remain the same, but the birds are not migrating as far south and may be disconnected from their winter food supplies.
5. As air temperatures rise, so too does the snow line in alpine areas. Animals living on or above the snow line will be forced into smaller areas. If they are unable to move to higher latitudes where the snow line is lower, it is inevitable that they will become extinct in their native ranges as they run out of food and space.

132. Global Warming and Effects on the Arctic (page 183)

1. Both low sea-ice albedo and area cause more heat to be absorbed by the land and sea. This heat causes the ocean's water to warm during the summer and therefore take longer to cool during the fall (autumn). This leads to winter sea-ice taking longer to form and being thinner than usual in the winter. Thin sea-ice has a lower albedo than thick sea-ice and melts faster in the spring leading to even less sea-ice the following winter. Thus thin ice and small area cover causes even thinner and less cover the following year in a potentially perpetual cycle until the sea-ice is lost.
2. Polar animals that live out on the sea-ice will be directly affected as the area covered by sea-ice reduces. Polar bears which hunt out on the ice find it harder to find food and must swim longer distances to firm ice. The reduction of sea-ice cover may also affect species that live below the ice as more light and heat will penetrate the waters and to deeper depths.

133. Ocean Acidification (page 184)

1. (a) Ocean acidification describes the decline in the pH of ocean waters towards a less alkaline state.
   (b) Ocean acidification makes it harder for shell making organisms to obtain the carbonate ions they need to build shells causing them to have thinner or deformed shells.
2. (a) Since the 1850s the pH of the oceans has been declining.
   (b) A rise in atmospheric CO_2 level. CO_2 dissolves into water forming carbonic acid, which lowers the pH.

134. Applying the Precautionary Principle (page 185)

1. The company planning to exploit the resource would have to show that their operation would not adversely affect the environment. If they could not, then the resource would not be allowed to be extracted.
2. Examples may include: flood control measures (e.g. use of stop banks and dams), genetically modified food (starvation in famine plagued countries), damming of rivers for hydroelectricity projects (use of more coal fired power stations), use of pesticides (e.g. organophosphates).

135. Chapter Review (page 186)

No model answer. Summary is the student's own.

136. KEY TERMS: Did You Get it? (page 188)

1. autotroph (M), carbon cycle (G), consumer (E), detritivore (J), ecological pyramid (I), food chain (A), food web (D), global warming (B), greenhouse gas (L), heterotroph (F), quadrat (C), saprotroph (H), trophic level (K)
2. Corn: producer
   Mouse: 1st order (primary) consumer
   Corn snake: 2nd order (secondary) consumer
   Hawk: 3rd order (tertiary) consumer
3. Student's own research and opinion. Students need to look at the validity and bias of the information and sources when formulating an opinion.

137. Genes, Inheritance, and Selection (page 191)

1. The only source of allelic variation in organisms that reproduce asexually is mutation.
2. Each individual has an effectively unique combination of alleles. Each individual is subjected to selection pressures that will affect its ability to survive and reproduce. In this way, the individual's allele combination is tested, with the better suited being able to produce more surviving offspring.
3. Sexual reproduction, through independent assortment and random combination of gametes at fertilization, shuffles alleles into new combinations. The new allele combinations provide variation, the raw material on which natural selection acts. Selection acts on the phenotype (being the result of genotype and environment). Selection acts for or against particular allele combinations, enhancing or reducing the representation of these in the next generation. Selection therefore sorts the variability in a gene pool and establishes adaptive genotypes.

138. The Fossil Record (page 192)

1. Preserved remains provide a record of the organisms alive when the sediment was laid down. Different stratigraphic levels record different organisms.
2. The evolution of the modern horse is well documented in the fossil record with numerous fossils, including transitional fossils, which show the major features of equid evolution over time (change in body size, limb length, tooth structure, toe reduction). The extensive fossil record has enabled scientists to put together a comprehensive and robust model of horse evolution (including speciation and extinctions).

139. Selection and Population Change (page 193)

1. (a) As milk yield in Holstein cows increases fertility decreases.
   (b) It suggests the genes for high milk yield and low fertility (or low milk yield and high fertility) are carried on the same chromosome (linked) and are very close together so that no crossing over and recombination occurs.
2. Milk yield will ultimately be limited by fertility (very high yield cows will have lower fertility and produce fewer offspring).
3. Selective breeding is an artificial form of natural selection usually based around the improvement of one or two traits desired by humans, in this case high milk yield. Natural selection favours traits that will increase reproductive success but this may result in a compromise (best balance) between correlated (genetically linked) traits, e.g. milk production and offspring production (fertility) are both costly in terms of energy expenditure so in nature there will be a balance. Humans have selected for milk yield over fertility to the detriment of the latter (probably because low fertility can be

ISBN: 978-1-927173-94-7

overcome with the use of administered hormones and other reproductive technologies).

140. Homologous Structures (page 194)

1. (b) Human arm: Modified for tree climbing (flexible joints) and improved dexterity of fingers.
   (c) Seal flipper: Modified to increase surface area and streamlined to function as a paddle.
   (d) Dog foot: Modified for swift running in pursuit of prey. Walks on toes, long limbs to provide lengthened, running stride.
   (e) Mole forelimb: Short and strong limb. Shovel-like paw with sharp claws for digging and propelling itself underground.
   (f) Bat wing: Modified into a wing for flying. Very long metacarpals and fingers stretch the skin into a wing.
2. In the forelimb wing of bats, the fingers (phalanges) are highly elongated to provide the structure for the skin that form the flight surface. In birds, the finger bones have been greatly reduced or lost with the fore arm providing the rigid front part of the wing so that most of the flight surface is made from feathers rather than skin.
3. The limbs all share the same basic bone anatomy, although highly modified in some cases. It is possible to match bone for bone but, at the same time, recognize how individual bones or bone groups have changed to better perform a new function for the animal in its new niche.

141. Divergence and Evolution (page 195)

1. (a) Sequential
   (b) Divergent
   (c) From common ancestor D: W, B, P, and H
   From common ancestor B: P and H
   (d) The trait was first developed in species B, then passed on to P and H.
2. (a) Divergence refers to the evolution of two (or more) species from a common ancestor (it is also called branching evolution). Adaptive radiation is a form of divergence where many species evolve from a common ancestor to occupy a range of new niches.
   (b) Divergent evolution involves branching of two or more species from a common ancestor. In contrast, sequential evolution involves evolutionary change in a single species over time without branching. In sequential evolution, one species accumulates changes gradually so that, over time, it becomes different enough to be identified and classified as a new species.

142. Mechanism of Natural Selection (page 196)

1. 1) Overproduction of the population, 2) genetic variation in the population, 3) competition for resources and survival of those with more favourable variations (natural selection), 4) inheritance of favourable variations and proliferation of individuals with these variations.
2. Mutations (creates new alleles) and sexual reproduction (produces new combinations of alleles).
3. The changes in the inherited characteristics of a population over many generations.
4. A population generally produces more offspring than will survive to reproduce. There is genetic variation in the offspring. Some of this variation will result in offspring phenotypes that are more suited to the prevailing environment than others. Competition will select for favourable phenotypes. The genetic component of the variation will be inherited by the next generation so that proportionally more of the favourable phenotype will be present in the next generation of offspring. Over time, the favourable phenotypes will proliferate and unfavourable phenotypes will become rare or disappear.

5.

Beetle population	% Brown beetle	% Red beetles	% Red beetles with spots
1	86.7	6.7	6.7
2	46.6	33.3	20
3	20	46.7	33.3

143. Adaptation (page 198)

1. Adaptive features are genetically determined traits (including physiological attributes) that have a function to the organism in its environment. Acclimatization refers to the changes made by an organism during its lifetime to environmental conditions.
2. Shorter extremities are associated with colder climates, whereas elongated extremities are associated with warmer climates. The differences are associated with heat conservation (shorter limbs/ears lose less heat to the environment).
3. A compact body size with a low surface area to volume ratio conserves more heat and has more heat producing mass relative to the surface area over which heat is lost.

144. Melanism in Insects (page 199)

1. The appearance of the wings and body (how speckled and how dark the pigmentation).
2. The selection pressure (the differential effect of selective predation on survival) changed from favouring the survival of light coloured forms in the unpolluted environments (prior to the Industrial Revolution) to favouring the dark morph (over the light morph) during the Industrial Revolution (when there was a lot of soot pollution). In more recent times, with air quality improving, the survival of the light coloured forms has once again improved.
3. As the frequency of the M allele increased so did the frequency of the dark form. Similarly as the frequency of the m allele decreased so did the grey morph.
4. The frequency of the darker form fell from 95% to 50%.

145. Selection for Beak Size in Darwin's Finches (page 200)

1. (a)

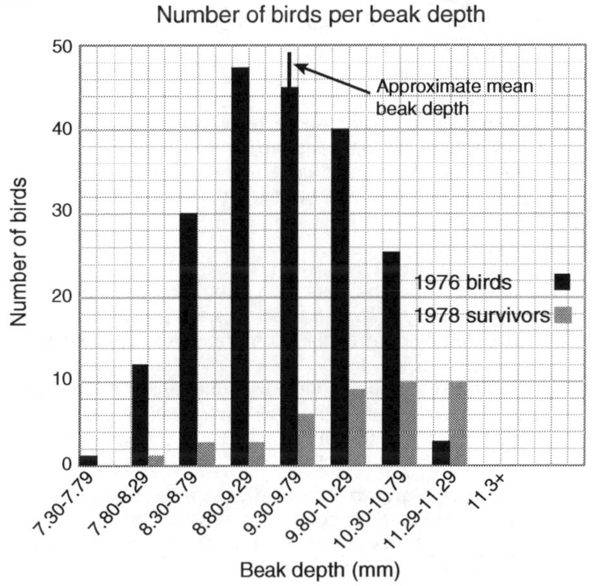

ISBN: 978-1-927173-94-7

(b)

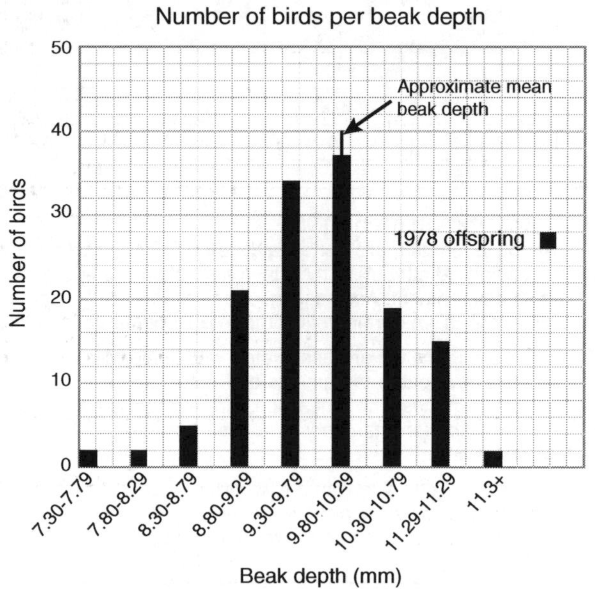

2. (a) See graphs above.
   (b) Approximately 0.5 mm
   (c) Yes, beak depth is heritable. If the drought continues the population may become increasingly dominated by individuals with deeper beaks..Smaller beak sizes will become increasingly rare or absent.
3. Smaller seeds were probably eaten first (as beaks were smaller). This left the birds competing for larger seeds such that birds with larger beaks were more successful and were more likely to survive.

146. The Evolution of Antibiotic Resistance (page 201)

1. Resistance can become widespread as a result of:
   (a) Transfer of genetic material between bacteria (horizontal evolution).
   (b) Increasing resistance with each generation a result of natural selection (vertical evolution). In this case, the antibiotic provides the selective environment.
2. (a) The genes for antibiotic resistance probably arose as a result of copying errors during DNA replication.
   (b) The genes remained in the bacterial genome because they were not detrimental to the bacteria, i.e. there was no selection against them.
   (c) Antibiotics are now widely used to treat bacterial infections, providing a strong new selective environment for bacteria. Having an antibiotic-resistant gene is an advantage, so the gene proliferates in the population. Because of the rapid generation time in bacteria, the gene can be rapidly spread through the population.

147. Investigating Evolution (page 202)

1. (a) Fitness varies between and within populations. The evolutionary path of a species is not predetermined, it is directed by current selection pressures (e.g. low glucose).
   (b) *E. coli* populations in the experiment were grown in a very limited set of conditions (low glucose). Over time, fitness in that particular environment increased. If placed in a different set of growth conditions, they may no longer have any advantage, and may be disadvantaged (lower fitness).
2. An increase in growth rate is advantageous because it allows *E.coli* to rapidly colonize a new environment or take advantage of a new set of environmental conditions, and thus out-compete other organisms.
3. The mutation at generation 15,000 is significant because it meant that for the first time *E.coli* could metabolize citrate, a component of the growth media. This provided an advantage over other populations and increased relative fitness.

148. The Evolution of Insecticide Resistance (page 203)

1. (a) and (b) any two in any order:
   – Insect populations tend to reproduce quickly (generation times are short) so the chance that a mutation conferring resistance will arise and be passed on is much greater than in populations that reproduce more slowly.
   – Partial resistance can arise at several levels (e.g. behavioural, mechanical, biochemical). Through sexual reproduction in a given selective environment, offspring can acquire a sufficient number of mechanisms to develop full resistance.
   – Applications of insecticide do not always reach their target in the correct dosage and so do not kill 100% of the population, allowing surviving individuals to reproduce and pass on even slight resistances.
2. The insecticide application may not kill off the entire insect population. Those that survive will pass on any resistance they have to the next generation. Periodic insecticide applications act as a selective agent by allowing only the most resistant insects to survive and pass on their genes.
3. Resistance to synthetic insecticides has a number of implications for biodiversity, health, and economics. In order to maintain the kill rate of the insecticide, farmers often increase the amount/potency/toxicity of insecticide applied. This leaves more residue on the plant and so is potentially more dangerous to humans. Crops may have to be withheld longer before going to market and so cause loss of income to the farmer. Increasing resistance among insect vectors of disease is also a problem for human populations (e.g. resistance in mosquito vector for malaria), reducing the options for controlling the spread of disease through human populations.

149. The New Tree of Life (page 204)

1. – Eukaryote groups are given much less prominence, reflecting the true diversity of the prokaryote groups.
   – Archaea are recognized as distinct to reflect their uniqueness and indicate their true relationship to eukaryotes and to other prokaryotes.

150. Classification System (page 205)

1.

(a)	(b)
1. Domain	1. Eukaryotes
2. Kingdom	2. Plantae
3. Phylum	3. Angiospermophyta
4. Class	4. Monocotyledoneae
5. Order	5. Fagales
6. Family (given)	6. Fagaceae
7. Genus	7. *Quercus*
8. Species	8. *robur*

2. (a) Binomial nomenclature
   (b) The genus and species (generic and specific name).
3. (a) and (b) in any order:
   (1) Avoid confusion over the use of common names for organisms, (2) provide a unique name for each type of organism, (3) attempt to determine/define the evolutionary relationship of organisms (phylogeny).
4. There are many possibilities. Common examples are:
   Keep **P**ond **C**lean **O**r **F**roggy **G**ets **S**ick
   Kids **P**laying **C**atch **O**n **F**reeways **G**et **S**mashed
5. Biochemical methods have helped identify lineages based on DNA and protein similarities (also see activities 149, 156, 157). In some cases this has caused a reorganization of a groups of lineages and in others it has confirmed the lineages already described. DNA sequences between related organisms can be compared for degree of similarity (more similar = more closely related). Amino acid sequences can be used in the same way. Molecular evidence is widely used to establish phylogenetic relationships in cladistics.

ISBN: 978-1-927173-94-7

152. Features of Plants (page 212)

1. **Bryophyta features**: Lack vascular tissues. Small and restricted to damp environments. Root-like rhizoids anchor plant in ground. In mosses, the plant body is leaf-like. In liverworts, it may be leaf-like or thalloid.
2. **Filicinophyta features**: True vascular plants. Plant body consists of underground stem or rhizome which bears the leaves (fronds) and roots. Conspicuous spore cases on the undersides of fronds. Clearly defined alternation of generations between large, leafy sporophyte and small, heart-shaped gametophyte.
3. **Coniferophyta features**: True vascular plants. Most are woody plants bearing seeds in cones. Other have naked seeds. Many are evergreen. Male cones shed pollen. Seeds are contained in the female cones.
4. **General Angiospermophyta features**: Vascular plants, producing flowers, fruits, and seeds.
   (a) Monocot features: Herbaceous (non-woody). Flower parts in multiples of three. Leaves narrow with parallel veins. Seeds have a single cotyledon.
   (b) Dicot features: Herbaceous or woody. Flower parts in multiples of four or five. Leaves often broader than in monocots with netted (branching) vein pattern. Seeds have two cotyledons.

153. Features of Animal Taxa (page 213)

1. **Porifera**: Lack organs and a nervous system. Possess spicules. The bodies consist of a non-living jelly-like mass between two main layers of cells.
2. **Cnidaria**: Body radially symmetrical. Medusoid (jellyfish) or hydroid (Hydra) body shape. Body wall of two layers separated by jellylike mesoglea. Tentacles with stinging cells for prey capture.
3. **Platyhelminthes**: Unsegmented body. No coelom. (internal body space). Mouth but no anus. No respiratory or circulatory system.
4. **Annelida**: Body more or less cylindrical and obviously segmented. Move by peristalsis, short appendages, or whole body undulations (leeches).
5. **Arthropoda**: Insects, crustaceans, arachnids, and myriopods are all arthropods and share some common features as follows; an exoskeleton of chitin and protein that is shed periodically to allow growth. Paired jointed appendages (legs, mouthparts). Segmented body but with a tendency for loss, fusion or specialization of segments to various degrees in different groups.
6. **Mollusca**: Unsegmented. Muscular foot for movement. Upper body forms a fleshy mantle which may secrete a protective calcium carbonate shell. Scraping feeding organ or radula. Otherwise the group is very diverse. Aquatic and terrestrial. Commonly recognized groups are the shelled gastropod snails and sea slugs and the bivalved molluscs (mussels etc.). The cephalopods (squid and octopus) are highly specialized, with a foot divided into tentacles and highly developed eyes.
7. **Echinodermata**: Marine. Adults radially symmetrical. Body wall rigid (with a skeleton of tiny calcareous plates). Spines are common. Some (sea lilies) are attached, most are mobile. Fluid filled tube feet used for locomotion. In most, mouth points down.
8. **Chordata**: Dorsal notochord, dorsal nerve cord, pharyngeal slits present, and post anal tail.
9. **Features of the sub-phylum Vertebrata**: Internal bony skeleton, paired nostrils and eyes.
   (a) **Chondrichthyes**: skeleton made of cartilage. No swim bladder, respiration occurs over exposed gill slits, placoid scales. Fertilization is internal.
   (b) **Osteichthyes**: skeleton made of bone. Swim bladders present in most species. Ganoid, cycloid, or cytenoid scales. Operculum (hard bony flap) covers the gills. Fertilization often external.
   (c) **Amphibia**: Adaptations for both aquatic and terrestrial life. Ectothermic. Smooth (non-scaly) skin covered with mucous glands. Moveable eyelids. Tympanic membrane for ear present in frogs. Muscular, protusible tongue. Legs and feet (may be webbed) adapted both for swimming and for moving on land.
   (d) **Reptilia**: Largely terrestrial or semi-aquatic. Dry, thick skin with horny scales or plates. May have legs or be legless (snakes). Tympanic membrane often present. Most lay shelled eggs. Most have well developed eyes.
   (e) **Aves**: Endotherms. Horny scales on feet, but feathers over most of the body. Beak, lacking teeth. Adapted for flight (wings, light skeleton) although some are secondarily flightless. Lay heavily shelled eggs.
   (f) **Mammalia**: Endotherms covered with hair or fur. Glandular skin. Young born live except monotremes which lay eggs. Mammary glands produce milk for young. Teeth often highly specialized. (Internal features: secondary palate and diaphragm).

154. Classification Keys (page 215)

1. The case (presence or absence and specific features of the case).
2. A *Oxyethira*
   B *Hudsonema*
   C *Olinga*
   D *Aoteapsyche*
   E *Hydrobiosis*
   F *Helicopsyche*
   G *Triplectides*
3. Insect order (common name)
   (a) Plecoptera (stoneflies)
   (b) Hemiptera (bugs)
   (c) Coleoptera (beetles)
   (d) Odonata (dragonflies and damselflies)
   (e) Lepidoptera (moths and butterflies)
   (f) Trichoptera (caddisflies)
   (g) Emphemoptera (mayflies)
   (h) Megaloptera (dobsonflies)
   (i) Diptera (true flies)

155. Keying Out Plant Species (page 217)

1. There are many different keys that students could have produced. The important idea is that each step clearly distinguishes between two groups and can be easily followed by someone else to get the same answer. An example of a written key given below:

   A dichotomous key to some maple species:

   1a Adult leaves with five lobes................2

   1b Adult leaves with three lobes 4

   2a Leaves 7.5-13 cm wide, with smooth edges, lacking serrations along the margin. U shaped sinuses between lobes: **Sugar maple**, *Acer saccharum*

   2b Leaves with serrations (fine teeth) along the margin.......3

   3a Leaves 5-13 cm wide and deeply lobed: **Japanese maple**, *Acer palmatum*

   3b Leaves 13-18 cm wide and deeply lobed: **Silver maple**, *Acer saccharinum*

   4a Leaves 5-15 cm wide with small sharp serrations on the margins. Distinctive V shaped sinuses between the lobes: **Red maple**, *Acer rubrum*

   4b Leaves 7.5-13 cm wide without serrations on the margins. Shallow sinuses between the lobes; **Black maple**, *Acer nigrum*

ISBN: 978-1-927173-94-7

156. Cladograms and Phylogenetic Trees (page 218)

1.

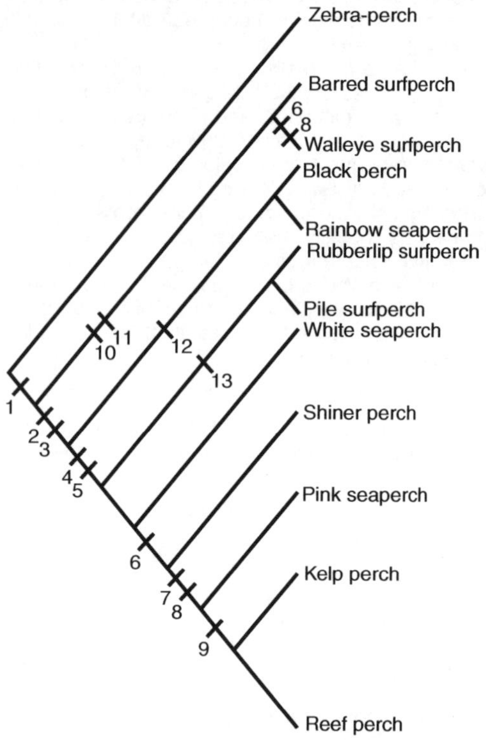

Teacher's should be aware that, while this is the most likely solution students will come up with (being the order presented in the table), other options are equally parsimonious and equally correct. Without more information, students cannot determine conclusively the final position of certain groups, e.g. kelp and reef perch could swap position, as could white seaperch and the rubberlip-pile branch. Rubberlip and pile surfperches could also swap positions. Ask your students what they would need to establish the definitive order of these groups (the answer being more characters).

2. (a) 6,8
   (b) A likely selection pressure is predation by other fish. This selects for a reduction in body size (6) as a way of being less obvious to predators. Prey fish take shelter at depth (away from light), selecting for larger eye size (8) to collect the reduced amount of light at greater depth.
3. You must make the assumption that the most parsimonious cladogram is the correct one.
4. A shared characteristic, as its named suggests, is shared by members of a group. It may show they are related. A shared derived characteristic is a trait that evolved in the ancestor and is shared by all its descendents.
5. Data match 5. The mutation from C to T must have occurred separately in pig and whale/hippo ancestors.
6. The phylogenetic tree can be tested by constructing trees based on a wide variety of characteristics (i.e. molecular and morphological) and/or using various phylogenetic methods. Trees obtained using different methodologies and different data are all likely to agree if the phylogeny is correct.

   Teacher's note: A typical phylogenetic analysis involves many decisions, including what type of data to sample, what phylogenetic method to apply, whether or not to order or weight characters, and which taxa and characters to include or exclude. Different choices can lead to very different trees. In practice, computer simulations, statistical analyses (such as bootstrap analysis) and agreement between trees from different data sets are all used to test proposed phylogenetic trees. Students should be aware that cladistic analysis is only one method for constructing a phylogenetic tree.

157. Cladistics (page 220)

1. (a) All members of the group have descended from a common ancestor (they are monophyletic).
   (b) A group is based on shared derived characteristics. The descendents of the ancestor share characteristics with the ancestor but may not share all their characteristics with other descendents.
2. Cladistics relies on shared characters that are derived from an ancestor and the order in which those characters were derived as opposed to similarities and differences between groups. Using cladistics, relationships between organisms will show an evolutionary history rather than simply a relationship in morphology.
3. (a) Biochemical evidence compares DNA and proteins between species and provides a more direct measure of common inheritance.
   (b) Biochemical evidence should not be used alone when the phylogenetic tree produced does not make biological sense (this depends heavily on the DNA sequence or protein chosen for the analysis). This is why care is taken to choose highly conserved genes and proteins for constructing phylogenies as differences are likely to biologically meaningful.
4. (a) Pongidae (b) Hominidae

158. Chapter Review (page 222)

No model answer. Summary is the student's own.

159. KEY TERMS: Did You Get it? (page 224)

1. adaptation (G), cladogram (B), dichotomous key (A), Eukaryotes (F), fossil record (I), genus (E), homologous (J), kingdom (C), natural selection (K), order (D), **phylogeny** (apologies for typographical error in workbook) (L), shared derived characteristic (H)
2. (a) A possible dichotomous key (student answers may vary):
   1a Length of all sides equal.....A
   1b Length of sides not equal 2

   2a Object has a distinct tail 3
   2b Object has no tail.....5

   3a Object has circular hole near curved end.....B
   3b Object has no hole.....4

   4a Object has tail split into two lobes.....F
   4b Object's tail is solid.....D

   5a Object has three straight sides and one end curved....C
   5b All sides are straight.....E

   (b)

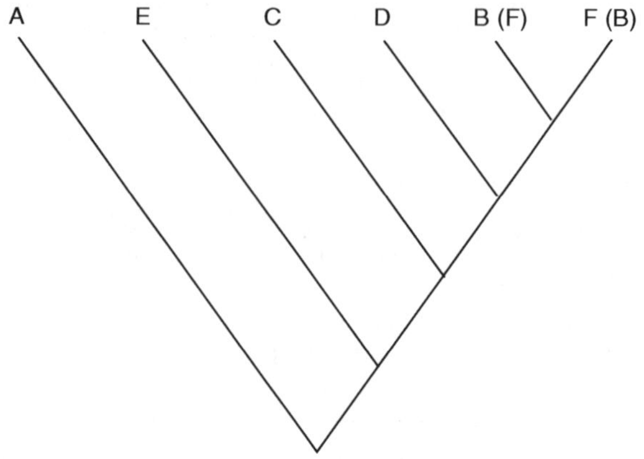

160. The Role of the Digestive System (page 227)

1. (a) Food is broken down mechanically (e.g. by chewing) and by chemical means (e.g. digestive enzymes).
   (b) Food must be broken down into smaller molecules so that it can be absorbed across the intestinal wall and utilized by the body.

ISBN: 978-1-927173-94-7

2. There are several different types of digestive enzymes, each hydrolyses (breaks down) different types of food. Proteases hydrolyse proteins and peptides into amino acids, Amylases hydrolyse carbohydrates, and lipases hydrolyse lipids.

161. Moving Food Through the Gut (page 228)

1. Contraction of circular muscle behind the food squeezes it through the gut. Contraction of longitudinal muscles in front of the food shortens and widens the gut to receive the food.
2. (1) Move food through the gut, (2) Mix fecal matter.
3.

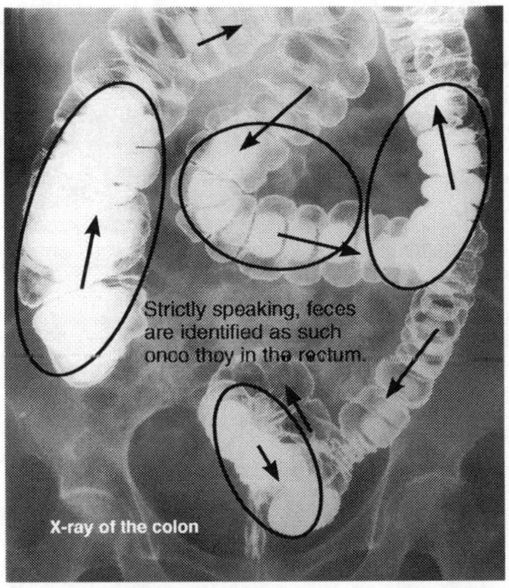

162. The Stomach (page 229)

1. Stomach has a role in storage and digestion - it secretes acid and a protein-digesting enzyme and mixes food.
2. HCl activates pepsinogen into its active form, pepsin. It also helps to kill any bacteria in food and denatures proteins.
3. Contraction of muscles in the stomach wall.

163. The Small Intestine (page 230)

1. (a) Duodenum, jejunum, ileum
   (b) Duodenum - most of the chemical digestion occurs here. Jejunum and ileum - most of the absorption occurs here.
2. Intestinal villi increase the surface area of the gut for absorption.
3. Enzymes are found in the pancreatic and intestinal juices, and in bile, which are all secreted into the duodenum. Enzymes are also bound to the surfaces of epithelial cells.
4. In general they act in alkaline conditions (pancreatic amylase acts in slightly acidic conditions).

164. The Large Intestine, Rectum, and Anus (page 231)

1. Main purpose of the large intestine is to absorb water and electrolytes.
2. Too little water absorption causes diarrhoea (watery feces), too much water absorption causes constipation (hard feces).

165. Digestion, Absorption, and Transport (page 232)

1. Amylase hydrolyses starch into the disaccharide maltose. The maltase enzyme then hydrolyses maltose into glucose. Glucose can then be absorbed across the intestine and utilized by the body.
2. The human digestive system does not produce enzymes capable of breaking cellulose down into its monomers.
3. Amylase produced in the salivary glands is denatured in the stomach. More amylase is secreted into the chyme by the pancreas once the food has passed the stomach.
4. (a) Active transport (e) Active transport
   (b) Facilitated diffusion (f) Diffusion
   (c) Active transport (g) Diffusion
   (d) Active transport (h) Diffusion
5. The nutrients are constantly transported away once they are absorbed into the blood.
6. (a) Having no starch in the water sets up a concentration gradient and simulates the intestinal-blood concentration environment in the body.
   (b) Large molecules cannot pass through the partially permeable membranes of cells. They must be broken down into smaller molecules first.

166. The Digestive Role of the Liver (page 234)

1. The liver produces bile and metabolizes fats, carbohydrates, and amino acids (especially with respect to converting glucose to glycogen and the metabolism of excess amino acids in the urea cycle). It stores iron, copper, and vitamins (absorbed from the food but not necessarily required immediately).
2. Glucose is transported to the liver in the bloodstream via the hepatic portal vein.
3. The liver stores glucose as glycogen when there is excess glucose in the blood and can release glucose into the blood when blood glucose levels drop. It can also produce glucose from non-carbohydrate sources.

167. Summary of the Human Digestive Tract (page 235)

1. Structures as follows:

   A Mouth and teeth
   B Salivary glands
   C Esophagus
   D Liver
   E Stomach
   F Pancreas
   G Gall bladder
   H Colon (or large intestine)
   I Small intestine
   J Rectum
   K Appendix
   L Anus

   Region responsible for each stated function as follows:
   (a) I - small intestine
   (b) J - rectum
   (c) H - colon
   (d) E - stomach
   (e) F - pancreas (or B)
   (f) D - liver
   (g) B - salivary gland
2. A. Longitudinal muscle C. Mucosa
   B. Circular muscle D. Epithelium
3.

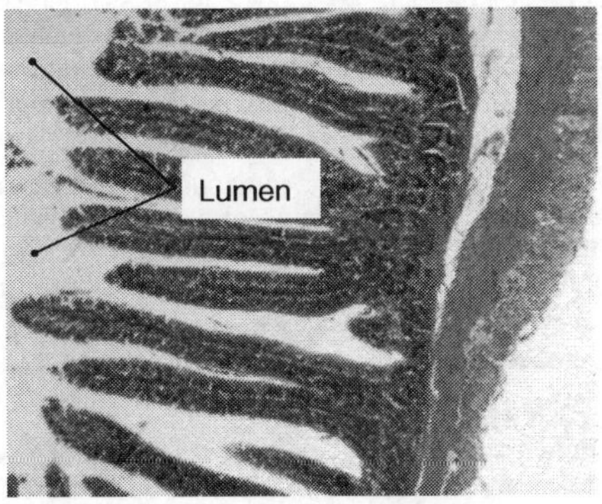

4. Circular muscle is viewed from the side, appearing as long fibres, while the longitudinal muscle has been cut transversely and so appears 'end on'.

168. The Circulatory System (page 236)

1. (a) Head (d) Gut (intestines)
   (b) Lungs (e) Kidneys
   (c) Liver (f) Genitals/lower body

ISBN: 978-1-927173-94-7

2. Circle pulmonary artery and pulmonary vein.

169. Blood Vessels (page 237)

1. (a) Arteries carry blood away from the heart to the capillaries within the tissues.
   (b) Capillaries exchange oxygen and nutrients in the blood with carbon dioxide and wastes from the cells.
   (c) The veins collect blood from the capillaries and return it to the heart.
2. The valves prevent blood flowing backwards in the veins. This prevents pooling in veins in the arms and legs and maintains the flow of blood towards the heart.
3. A: Vein. There is only a thin layer of muscle and the lumen is relatively large
   B: Artery. A thick layer of muscle can be seen in the photograph. The wall is thicker and the lumen smaller.
4. (a) Veins have less elastic and muscle tissue than arteries.
   (b) Veins have a larger lumen than arteries.
5. Most of the structural differences between arteries and veins are related to the different blood pressures inside the vessels. Blood in veins travels at low pressure and veins do not need to be as strong, hence the thinner layers of muscle and elastic tissue and the relatively larger lumen. Teacher's note: There is still enough elastic and muscle tissue to enable the veins to adjust to changes in blood volume and pressure.
6. (a) Capillaries are very small blood vessels forming networks or beds that penetrate all parts of the body. The only tissue present is an endothelium of squamous epithelial cells. In contrast, arteries have a thin endothelium, a central layer of elastic tissue and smooth muscle and a thick outer layer of elastic and connective tissue. Veins have a thin endothelium, a central layer of elastic and muscle tissue and a thin outer layer of elastic connective tissue. In addition, veins also have valves.
   (b) Capillaries are not under any pressure from blood flow, but must be thin to allow efficient exchange of nutrients and wastes between the blood and tissue. Note: Arteries must be strong enough to withstand the high flow and pressure of blood pumped from the heart. Blood returning to the heart through veins is at a lower pressure, so less elastic and muscle tissue is required, but they contain valves to prevent blood backflow.
7. In blood vessels, most of the resistance opposing blood flow is due to vessel diameter. As vessel diameter decreases, the resistance increases and blood flow decreases. The slow blood flow through capillaries allows sufficient time for exchange of gases and nutrients.
8. Venous blood oozes out in an even flow from a wound because it has lost a lot of pressure after passing through the narrow capillary vessels (with their high resistance to flow). Arterial blood spurts out rapidly because it is being pumped directly from the heart and has not yet entered the capillary networks.
9. Capillaries form dense networks in tissues with high metabolic rates to ensure sufficient exchange of gases and quick removal of wastes. This enables high metabolic rate to be maintained without the build up of toxic wastes or reduced supply of oxygen and nutrients.

170. Capillary Networks (page 239)

1. Capillary networks are branching networks of fine blood vessels where exchanges between the blood and tissues take place. Blood enters the network at the arteriolar end and is collected by venues at the venous end. The true capillaries form a network outside of the vascular shunt.
2. The smooth muscle sphincters regulate the blood flow to the capillary network. They contract to restrict blood flow to the network and relax to allow blood to flow in. The vascular shunt connects the arteriole and venule and allows blood to bypass the capillaries when the sphincters are contracted.
3 (a) Situation A would occur when the body is restricting blood flow to the capillaries, for example when trying to conserve heat by diverting blood away from the extremities.
   (b) Situation B would occur when the body is trying to remove excess heat by diverting blood to the skin and extremities or when the body is trying to provide extra blood to areas of high metabolism, e.g. when exercising or digesting food.
4. A portal venous system drains blood from one capillary network into another. An example is the hepatic portal system which drains blood from the capillary network in the gut to the capillary network in the liver. Normally capillary networks drain into veins that return directly to the heart.

171. The Heart (page 240)

1. Schematic of human heart. Labels (a)-(h)

(a) Pulmonary artery	(e) Aorta
(b) Vena cava	(f) Pulmonary vein
(c) Right atrium	(g) Left atrium
(d) Right ventricle	(h) Left ventricle

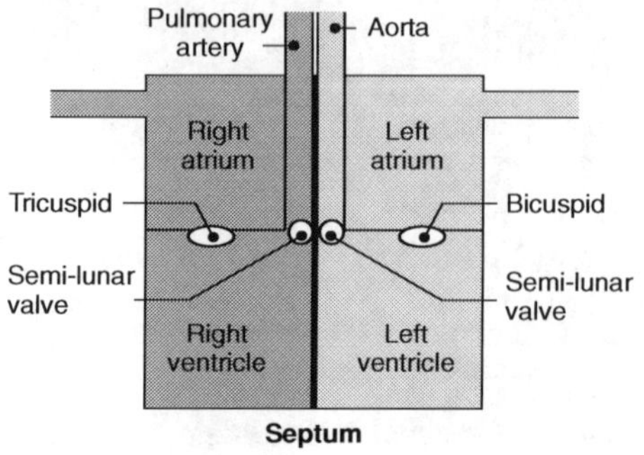

2. Valves prevent the back flow of blood into the atria and ensure blood flows in the correct direction.
3. (a) The heart has its own coronary blood supply to meet the high oxygen demands of the heart tissue.
   (b) There must be a system within the heart muscle itself to return deoxygenated blood and waste products of metabolism back to the right atrium.
4. If blood flow to a particular part of the heart is restricted or blocked (because of blocked blood vessel), the part of the heart muscle supplied by that vessel will die, leading to a heart attack or infarction.
5. A: arterioles B: venules
   C: arterioles D: capillaries
6. (a) The pulmonary circuit must operate at a lower pressure than the systemic circuit to prevent fluid accumulation in the lungs. The systemic circuit must also develop enough pressure to enable blood flow to increase to the muscles when required and to maintain kidney filtration rates without compromising blood supply to the brain.
   (b) The left ventricle is thicker to enable the left side of the heart to develop higher pressure. There is no such requirement of the right side, which pumps only to the low-pressure pulmonary circuit.
7. You are recording expansion and recoil of the artery that occurs with each contraction of the left ventricle.

172. Control of Heart Activity (page 242)

1. (a) Sinoatrial node: Initiates the cardiac cycle through the spontaneous generation of action potentials.
   (b) Atrioventricular node: Delays the impulse.
   (c) Bundle of His: Distributes the action potentials over the ventricles (resulting in ventricular contraction).
   (d) Intercalated discs: Electrical junctions that allow the impulses to spread very rapidly through the heart muscle.
2. Delaying the impulse at the AVN allows time for atrial contraction to finish before the ventricles contract.
3. Influencing heart rate via the CNS allows the heart to deliver

ISBN: 978-1-927173-94-7

blood efficiently in response to demand (e.g. exercise, fight or flight).

4. (a) Epinephrine increases heart rate
   (b) Norepinephrine

173. The Cardiac Cycle (page 243)

1. (a) Between the S and T section of the trace
   (b) The QRS complex (specifically R)
   (c) Ventricles are relaxed and filling (ventricular volume increasing).
2. During electrical recovery the heart muscle cannot contract so the heart has an enforced rest and will not fatigue, nor accumulate lactic acid.
3.

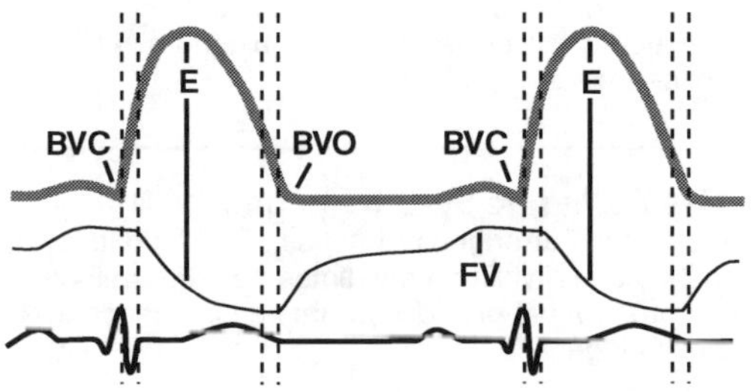

174. Review of the Heart (page 244)

1. (A) Epinephrine
   (B) Acetylcholine
   (1) Right atrium
   (2) Sinoatrial node (SAN or pacemaker)
   (3) Atrioventricular node
   (4) Bundle of His
   (5) Right ventricle
   (6) Aorta
   (7) Pulmonary artery
   (8) Left atrium
   (9) Purkyne fibres
   (10)Left ventricle
2. A = Contraction of the atria.
   B = Contraction of the ventricles.
   C = Relaxation and filling of ventricles.

175. Dissecting a Mammalian Heart (page 245)

1.

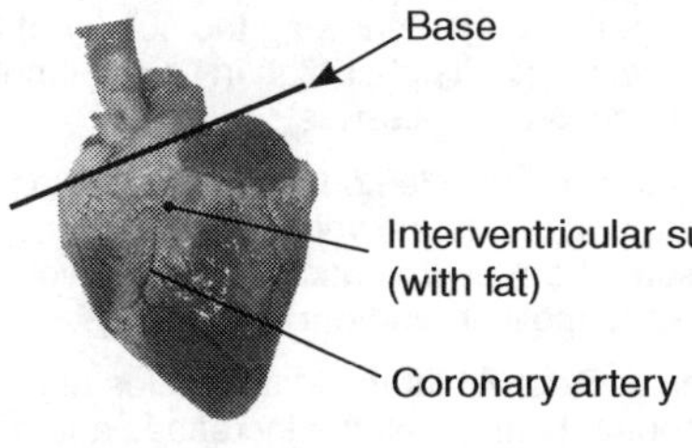

2. Vena cava
3. Aorta
4. Pulmonary artery
5. Chordae tendineae hold the valves between the atrium and ventricle closed during contraction of the ventricle.
6. The thickness of the ventricle walls (the right ventricle wall is relatively thin while the left ventricle wall is much thicker and very muscular).

176. Coronary Occlusions (page 247)

1. (a) Coronary occlusions are blockages of the coronary arteries.
   (b) The occlusion may completely block the blood flow or the plaque may break off and cause a blood clot which can block the artery.
2. High blood pressure, high levels of LDL cholesterol, smoking, and high blood sugar levels.

177. The Body's Defences (page 248)

1. The skin provides a physical barrier to prevent entry of pathogens. Skin secretions (serum and sweat) contain antimicrobial chemicals which inhibit microbial growth.
2. (a) Phospholipases kill bacteria by hydrolysing the phospholipids in cell walls and membranes.
   (b) Cilia move microbes, which are trapped in mucus, towards the mouth and nostrils, to be expelled.
   (c) Sebum has antimicrobial activity and (with sweat) a pH that is unfavourable for microbial growth.
3. Antibodies are proteins produced in response to antigens; they recognize and bind antigens. Antigens are foreign substances (often proteins) that promote the formation of antibodies (invoke an immune response).
4. (a) Phagocytosis destroys pathogens directly by engulfing them.
   (b) Antimicrobial substances (e.g. interferon) prevent multiplication of microbes (especially viruses).
   (c) Antibodies are produced against specific pathogens, and bind and destroy pathogens or their toxins.
5. A three tiered (hierarchical) system of defence provides a series of back-ups in case a pathogen breaches earlier barriers. Most microbes are excluded by the first line of defence, but those that penetrate the skin will usually be destroyed by white blood cells and the chemicals associated with inflammation. Failing this, the body will mount a targeted specific defence against the identified pathogen still remaining.

178. Blood Clotting and Defence (page 250)

1. Blood clotting prevents bleeding and invasion of bacteria by formation of a scab.
2. (a) Injury exposes collagen fibres to the blood.
   (b) Chemicals make the surrounding platelets sticky.
   (c) Clumping forms an immediate plug of platelets preventing blood loss.
   (d) Fibrin clot traps red blood cells and reinforces the seal against blood loss.
3. (a) Clotting factors catalyse the conversion of prothrombin to thrombin, the active enzyme that catalyses the production of fibrin.
   (b) If clotting factors were always present, clotting could not be contained; blood would clot when it shouldn't.

179. The Action of Phagocytes (page 251)

1. Foreign bodies and detected and engulfed. A phagosome forms around the foreign objects and fuses with a lysosome, which contains enzymes that digest the foreign body.
2. Phagocytes act against any foreign bodies rather than specific (identified) ones.
3. By looking at the ratio of white blood cells to red blood cells (not involved in the immune response). An elevated white blood cell count (specifically a high neutrophil count) indicates microbial infection.

180. Inflammation (page 252)

1. (a) Increased diameter and permeability of blood vessels. Role: Increases blood flow and delivery of leukocytes to the area. Aids removal of destroyed microbes or their toxins. Allows defensive substances to leak into the tissue spaces.
   (b) Phagocyte migration and phagocytosis. Role: To directly attack and destroy invading microbes and foreign substances.
   (c) Tissue repair. Role: Replaces damaged cells and tissues,

ISBN: 978-1-927173-94-7

restoring the integrity of the area.

2. Ability to squeeze through capillary walls (amoeboid movement). Ability to engulf material by phagocytosis.
3. Histamines and prostaglandins attract phagocytes to the site of infection.
4. Inflammation is not a response to a specific foreign body but to infection generally.

181. Antibiotics (page 253)

1. Antibiotics target the structures and metabolic pathways of bacteria, which are different to eukaryotic cells.
2. Bacteriostatic: chloramphenicol. The bacterial population remains the same over the length of the experiment.

   Bactericidal: ampicillin. The bacterial population decreases over the length of the experiment.
3. (a) Antibiotic resistance is the ability of bacteria to resist the actions of antibiotics and continue to reproduce.
   (b) When bacteria acquire resistance to several antibiotics it makes infections very difficult to treat. The infection may become worse and eventually kill the patient. MRSA is an example of a bacterium that has acquired this kind of multiple resistance.
4. The experiment demonstrated how to treat bacterial infections in a relatively simple way and led to the development of the first mass produced antibiotic treatments.
5. (a) A
   (b) 3
   (c) This concentration in most effective as it produces the best result with the least amount of antibiotic used.

182. Viral Diseases (page 255)

1. A highly infectious intracellular pathogen that replicates only inside the living cells of organisms.
2. Antibiotic are only effective against aspects of bacterial metabolism.
3. Collaboration between countries and organizations ensures diseases are monitored and any outbreaks are quickly contained. Heightened surveillance, reporting vaccination programmes, and fast communication ensure infectious diseases remain at low levels in the population.

183. HIV and AIDS (page 256)

1. (a) HIV infects helper T-cell lymphocytes.
   (b) HIV infects and kills the cells of the immune system that normally defend the body from infection. This leaves the body vulnerable to invasion by microbes that would not normally infect a healthy person.
2. (a) The virus rapidly increases in numbers within the first year of infection, followed by a large drop off in numbers in the second year. Over the next 3-10 years, the HIV population gradually increases again.
   (b) The helper T cell numbers respond to the initial infection by increasing in numbers. After about a year, their numbers steadily decrease as they are attacked and destroyed by the HIV.
3. (a) Transmission of HIV, any three of: Blood or blood products, vaginal secretions, breast milk, across the placenta, shared needles among intravenous drug users (contaminated with blood from other drug users), sexual intercourse: both homosexual (between males) and heterosexual (between men and women).
   (b) Blood products are screened for HIV antibodies. Any blood with a positive result is rejected for use.
4. The use of several different drug types at the same time attacks HIV at each stage of its life cycle giving the best chance of controlling the virus.
5. HIV has very high mutation rates. It can quickly acquire mutations that make it resistant to a particular antiviral drug. Incorrect use of antiviral drugs (such as only using one drug at a time) increases the risk that HIV will evolve a resistance to the drug. This is a problem as it allows the disease more opportunity to spread in the population as antiviral drugs become less effective.

184. Introduction to Gas Exchange (page 258)

1. Gas exchange is the way in which respiratory gases (oxygen and carbon dioxide) are exchanged with the environment. .
2. (a) Oxygen and carbon dioxide. (b) Diffusion.
3. A gas exchange surface provides a surface across which gases can diffuse, enabling exchanges with the environment.
4. (a) Large surface area to provide for a large amount of gas exchange (to meet the organism's needs).
   (b) Thin membrane that does not present a large barrier to diffusion of gases.
   (c) Associated with mechanisms (e.g. blood flow) to maintain the concentration gradient.

185. The Gas Exchange System (page 259)

1. (a) The structural arrangement (lobes, each with its own bronchus and dividing many times before terminating in numerous alveoli) provides an immense surface area for gas exchange.
   (b) The alveoli.
2. Respiratory (gas exchange) membrane is the layered junction between the alveolar cells, the endothelial cells of the capillaries, and their associated basement membranes. It provides a surface across which gases can freely move.
3. Cartilage supports the larger airways and keeps them open for air flow.
4. Type I pneumocytes contribute to the gas exchange membrane. They make up 90-95% of alveolar cells. Type II pneumocytes secrete surfactant, which decreases surface tension within the alveoli.
5. The surfactant reduces the surface tension of the lung tissue and counteracts the tendency of the alveoli to recoil inward and stick together after each expiration.
6. The lack of surfactant and high surface tension in the alveoli result in the collapse of the lungs to an uninflated state after each breath. Breathing is difficult and laboured, oxygen delivery is inadequate and, if untreated, death usually follows in a few hours.

186. Breathing (page 261)

1. Breathing ventilates the lungs, renewing the supply of fresh (high oxygen) air while expelling air high in CO_2 (gained as a result of gas exchanges in the tissues).
2. Breathing is the result of muscle contraction and relaxation that increases and decreases the volume of the thoracic cavity. The pressure changes that accompany the volume changes cause air to move in and out of the lungs.
3. (a) Quiet breathing: External intercostal muscles and diaphragm contract. Lung volume increases and air flows into the lungs (inspiration). Expiration occurs through elastic recoil of the ribcage and lung tissue (air flows passively out to equalize with outside air pressure).
   (b) In active breathing, muscular contraction is involved in both inspiration and expiration (expiration is not passive).
4. (a) External intercostals and diaphragm.
   (b) Internal intercostals and abdominal muscles
5. (a) Internal intercostals and abdominal muscles
   (b) External intercostals and diaphragm.
6. Antagonistic muscles bring about the changes in thoracic volume that enable air to be moved in and out of the lungs. The muscles for inspiration are in opposition to those for expiration. When one set is contracting, the other is relaxed.

ISBN: 978-1-927173-94-7

187. Measuring Lung Function (page 262)

1. (a) Taller people generally have larger lung volumes and capacities.
   (b) Males have larger lung volumes and capacities than females.
   (c) After adulthood, lung volume and capacity declines with age. Children have smaller lung volumes and lung capacities than adults.
2. (a) Forced volume is a more useful indicator of impairment of lung function than a tidal volume because people use only a small proportion of their lung volume in normal breathing.
   (b) Spirometry can be used to measure the extent of recovery of lung function after treatment.
3. (a) Tidal volume vol: 0.5 L
   (b) Expiratory reserve volume vol: 1.0 L
   (c) Residual volume vol: 1.2 L
   (d) Inspiratory capacity vol: 3.8 L
   (e) Vital capacity vol: 4.8 L
   (f) Total lung capacity vol: 6.0 L
4. G: Tidal volume is increasing as a result of exercise.
5. PV: $15 \times 0.4 = 6$ L
6. (a) During strenuous exercise, PV increases markedly.
   (b) Increased PV is achieved as a result of an increase in both breathing rate and tidal volume.
7. (a) There is 90X more CO_2 in exhaled air than in inhaled air ($3.6 \div 0.04$).
   (b) The CO_2 is the product of cellular respiration in the tissues. Note: Some texts give a value of 4.0% for exhaled air (100X the CO_2 content of inhaled air).
   (c) The dead space air is not involved in gas exchange therefore retains a higher oxygen content than the air that leaves the alveoli air. This raises the oxygen content of the expired air.

188. Exercise and Breathing (page 264)

1. (a) The graph produced is dependent on student's own response to exercise.
   (b) Students should see an increase in both heart rate and breathing rate during the exercise period.
2. (a) After one minute of rest, students should see a decrease in both heart rate and breathing rate. After five minutes both should have decreased even further, and may have returned to pre-exercise levels.
   (b) Once exercise is completed, the body's metabolic rate falls. The demand for energy and oxygen falls, and the heart rate and breathing rate will fall accordingly.

189. Smoking and Lung Cancer (page 265)

1. The graph clearly shows that smokers have a much higher fatality rate from lung cancer than non-smokers.
2. Smoking destroys the walls of the alveoli and reduces surface area for gas exchange. It also causes the production of extra mucus which can slow gas exchange.

190. Emphysema (page 266)

1. The economic impact of tobacco smoking-related diseases is two fold. Directly there are the considerable health costs associated with treating the consequences of diseases such as chronic bronchitis and emphysema, which often require repeated and prolonged hospital visits. Secondly, there is the economic cost of lost work days (estimated 24 million lost work days per annum).
2. This response is largely a student's own and there is no 'right' answer. Tobacco-related diseases generally have a detrimental effect on a person's quality of life and on the quality of life for those who end up as carers for debilitated relatives. There are additional associated health costs with hospital and doctor's visits and treatments, and the stress of living with a chronic condition which leaves the sufferer vulnerable to more serious lung infections. Personal testimonials, such as that from Jenny's daughter, provide a first hand account of what the personal costs can be and may give some people enough reason to reconsider their smoking habit or to stop them ever taking it up.

191. Nervous Regulatory Systems (page 267)

1. (a) Sensory receptors receive sensory information from (and about) the environment and respond by generating an electrical response (message).
   (b) Central nervous system (CNS) processes the sensory input and coordinates an appropriate response (through motor output).
   (c) Effectors bring about an appropriate response (a response to maintain homeostasis).
2. Nervous control is rapid. It acts in the short term and its effects are short lived. In contrast, hormonal control is slower and it is longer acting.

192. Neuron Structure and Function (page 268)

1. A neuron processes and transmits information as electrical impulses from the point of stimulus to the point of reception.
2. A larger axon diameter and increased myelination.
3. Myolin prevents ions from entering or leaving the axon along myelinated segments and so prevents leakage of charge across the neuron membrane. The current is carried in the cytoplasm so that the action potential at one node (gap in the sheath) is sufficient to trigger an action potential at the next. Myelin has another advantage too in reducing energy expenditure since fewer ions overall need to be pumped to restore resting potential after an action potential has passed.
4. Faster conduction speeds enable more rapid responses to stimuli.

193. The Nerve Impulse (page 269)

1. An action potential is a self-regenerating depolarization (electrochemical signal) that allows excitable cells (such as muscle and nerve cells) to carry a signal over a (varying) distance
2. (a) The nerve impulse "jumps" from node to node as depolarization at one node triggers an action potential that depolarizes the next node.
   (b) It increases conduction speed.
3. Letters should appear in the boxes in this order (top to bottom): B, D, E, C, A

194. Neurotransmitters (page 271)

1. To transmit signals from one neuron to the next across the synapse.
2. (a) Acetylcholine
   (b) The neurotransmitter took time to move through the solution from one heart to the next, causing the delay.
3. Neonicotoids mimic the action of acetylcholine in synapses. They bind irreversibly to the postsynaptic nicotinic acetylcholine receptors causing over-stimulation of the neuron.

195. Chemical Synapses (page 272)

1. A synapse is a junction between the end of one axon and the dendrite or cell body of a receiving neuron. Note: A synapse can also occur between the end of one axon and a muscle cell (neuromuscular junction).
2. Arrival of a nerve impulse at the end of the axon causes an influx of calcium. This induces the vesicles to release their neurotransmitter into the cleft.
3. Delay at the synapse is caused by the time it takes for the neurotransmitter to diffuse across the synaptic cleft.
4. (a) Neurotransmitter (NT) is degraded into component molecules by enzyme activity on the membrane of the

ISBN: 978-1-927173-94-7

receiving neuron (in this case, acetylcholinesterase (AChE) acts on ACh to produce acetyl and choline).
(b) The neurotransmitter must be deactivated so that it does not continue to stimulate the receiving neuron (continued stimulation would lead to depletion of neurotransmitter and fatigue of the nerve). Deactivation allows recovery of the neuron so that it can respond to further impulses.

5. The amount of neurotransmitter released influences the response of the receiving cell (response strength is proportional to amount of neurotransmitter released).

---

196. Chemical Imbalances in the Brain (page 273)

1. A malfunction in production or response to a neurotransmitter (e.g. serotonin) in the brain will affect transmission of signals in regions of the brain that control mood, emotion, and other behaviours. The imbalances and resulting disturbances are involved in mental illnesses such as depression and anxiety.
2. (a) Parkinson's disease is caused by reduced dopamine production due to the loss of nerve cells in the substantia nigra region of the brain. This reduces neural activity in the motor cortex and results in slow physical movement and uncontrollable tremors. It is treated with L-dopa, a dopamine precursor, which can cross the blood-brain barrier to the brain and produce more dopamine, relieving symptoms of the disease.
(b) Depression is caused by a reduction in serotonin released from the raphe nuclei, resulting in reduced stimulation of neurons in the serotonin pathways, which are related to mood and emotion. The deficiency can be corrected with drugs that prevent serotonin breakdown (MAOI) or prevent its re-uptake by presynaptic cells (thereby raising available levels in the brain).

---

197. Hormonal Regulatory Systems (page 274)

1. (a) Antagonistic hormones: Two hormones that have contrasting (counteracting) effects on metabolism. Examples: insulin and glucagon, parathormone (increases blood calcium) and calcitonin (lowers blood calcium).
(b) In general principle, the product of a series of (hormone controlled) reactions controls its own production by turning off the pathway when it reaches a certain level. If there is too little of the product, its production is switched on again.
2. Only the target cells have the appropriate receptors on the membrane to respond to the hormone. Other (non-target) cells will not be affected.
3. (a) Hormones must circulate in the blood to reach the target cells and a metabolic response must be initiated. This takes some time.
(b) Hormones bring about a metabolic change and often start a sequence of cascading, interdependent events, which take some time to conclude. Nervous responses continue only for the time that the stimulation continues.

---

198. Principles of Homeostasis (page 275)

1. Homeostasis is the relatively constant internal state of an organism, even when the external environment is changing.
2. Negative feedback mechanisms are self-correcting (the response acts to oppose changes to the input) so that fluctuations are dampened. This stabilizes physiological systems against excessive change and enables maintenance of a steady state.

---

199. The Endocrine System (page 276)

1. Hormone: Insulin
Secreted by: β cells of the pancreas
Role: Acts to lower blood glucose levels

Hormone: Leptin
Secreted by: Adipose tissue
Role: Inhibits appetite

Hormone: Estrogen, progesterone
Secreted by: Ovaries
Role: Control development of primary and secondary sex characteristics, maintain femaleness, stimulate the menstrual cycle, maintain pregnancy, and prepare the mammary glands for lactation.

Hormone: Melatonin
Secreted by: Pineal gland
Role: Regulates the sleep-wake cycle. Coordinates reproductive hormones.

Hormone: Testosterone
Secreted by: Testes
Role: Controls development of male genitalia, maintains maleness, and promotes sperm production.

Hormone: Glucagon
Secreted by: α cells in the pancreas
Role: Increase blood glucose levels

Hormone: Thyroxin
Secreted by: Thyroid gland
Role: Stimulates metabolism and growth and is involved in temperature regulation.

2 (a) Long distance travel disrupts the sleep-wake cycle by causing the body clock to be out-of-sync with the usual environmental light-dark cues.
(b) By taking melatonin when they reach their destination, travelers can sleep and also reset their biological clock to the new time zone.

3. Leptin is an appetite suppressing hormone. It acts on receptors in the hypothalamus, regulating appetite and food intake. A low leptin production, or resistance to its effects, can lead to over eating and obesity.
4. They have developed leptin resistance. The receptors no longer respond to leptin and appetite is not suppressed as appropriate when it should be.

---

200. Control of Blood Glucose (page 278)

1. (a) Stimulus: Rise in the levels of glucose in the blood above a set level (about 5.6 mmol L^{-1}).
(b) Stimulus: Fall in blood glucose levels below a set level (about 3.9 mmol L^{-1}).
(c) Glucagon brings about the production (and subsequent release) of glucose from the liver by the breakdown of glycogen and the synthesis of glucose from amino acids.
(d) Insulin increases glucose uptake by cells and brings about production of glycogen and fat from glucose in the liver.
2. Fluctuations in blood glucose (BG) and blood insulin levels are closely aligned. Following a meal, BG rises sharply and there is a corresponding increase in blood insulin, which promotes cellular glucose uptake and a subsequent fall in BG. This pattern is repeated after each meal, with the evening meal followed by a gradual decline in BG and insulin over the sleep (fasting) period. Negative feedback mechanisms prevent excessive fluctuations in blood glucose (BG) throughout the 24 hour period.
3. Humoral.

---

201. Diabetes Mellitus (page 279)

1. Diabetes causes high blood sugar because the cells cannot take up glucose due to difficulties with either insulin production or insulin responsiveness. The glucose remains circulating in the blood.
2. In type 1 diabetes, the beta cells of the pancreatic islets are destroyed by the body's immune system, and no insulin is produced. In type 2 diabetes the beta cells of the pancreatic islets still produce insulin but the quantities are insufficient or the cells no longer react to it. Causes of type 2 diabetes are often lifestyle factors such as obesity, high blood pressure, and lack of exercise. Treatment for type 1 diabetes involves regular insulin injections and dietary management. Type 2 diabetes is usually treated by changing the causative lifestyle factors (e.g. increasing activity, losing weight, and improving diet). Sometimes drugs and insulin may also be required.

ISBN: 978-1-927173-94-7

202. The Male Reproductive System (page 280)

1. (a) Vas deferens (b) Bladder (c) Seminal vesicle (d) Prostate gland (e) Epididymis (f) Urethra (g) Testis (h) Scrotal sac (scrotum)

2. (a) Conducts sperm to the urethra.
   (b) Stores urine.
   (c) Produces secretions that add to the semen.
   (d) Produces secretions that form part of the semen.
   (e) Coiled tube where sperm mature.
   (f) Conducts sperm and urine to the outside.
   (g) Produces sperm and testosterone (male hormone).
   (h) Holds the testes (outside the body at the cooler temperature required for sperm production).

3. Testosterone is involved in the development of the male genitalia in the embryo (primary sexual characteristics). At puberty, testosterone causes the secondary sexual characteristics to develop (sperm, body hair, muscle growth, deepening of voice). In adults it maintains sex drive and production of sperm.

4. Sex is determined by the presence or absence of the Y chromosome. XX produces a female. XY produces a male.

203. The Female Reproductive System (page 281)

1. (a) Fallopian tube (oviduct) (b) Ovary (c) Uterus (d) Cervix (e) Vagina (f) Clitoris

2. (a) The usual place of fertilization and also conducts the egg to the uterus.
   (b) Produces ova (eggs) and female reproductive hormones (estrogen and progesterone).
   (c) Houses developing embryo, establishes placenta.
   (d) Entry to uterus. Its mucus protects the uterus from invasion by microorganisms. It is a powerful muscle that prevents the loss of the fetus in pregnancy.
   (e) Birth canal. Receives penis in sexual intercourse, pathway for delivery of the baby at term during labour.

3. (a) A = Ovary
   (b) Ovulation
   (c) Oogenesis (gametogenesis is also correct).

4. Fertilization occurs in the Fallopian tube (oviduct).

204. The Menstrual Cycle (page 282)

1. (a) FSH (b) LH (surge)

2. (a) Estrogen
   (b) Promotes repair and growth of the uterine lining.
   (c) The follicles are reabsorbed.

3. (a) Progesterone (estrogen in smaller amounts).
   (b) Promotes full development of the endometrium. If fertilization occurs, progesterone maintains the thickened endometrium and prevents further follicle development.

4. Menstruation is triggered by a sharp drop in progesterone levels following degeneration of the corpus luteum.

205. Using Hormones to Treat Infertility (page 283)

1. Drugs are given to suppress ovulation. Injections of luteinizing hormone and follicle stimulating hormone are given daily to 10-14 days to stimulate the production of a larger number of eggs than usual. An injection of gonadotropin is given to trigger final maturation of the eggs. Eggs are harvested and mixed with sperm. Fertilized eggs are implanted in the uterus.

2. (a) LH: Stimulates egg production from the ovaries.
   (b) FSH: Stimulates egg production from the ovaries.
   (c) hCG: Triggers maturation of the eggs.
   (d) Progesterone: Supports uterine lining and maintains early pregnancy.

3. Ovulation is suppressed to prevent eggs being released too soon, making harvesting more successful.

206. Chapter Review (page 284)

No model answer. Summary is the student's own.

207. KEY TERMS: Did You Get it? (page 286)

1. (a) Peristalsis (b) Small intestine (c) To increase absorption area (d) The pancreas

2. (a) Artery (b) Vein (c) Capillary

3. (a) Platelets and blood proteins
   (b) Leucocytes or white blood cells
   (c) Bacteria

4.

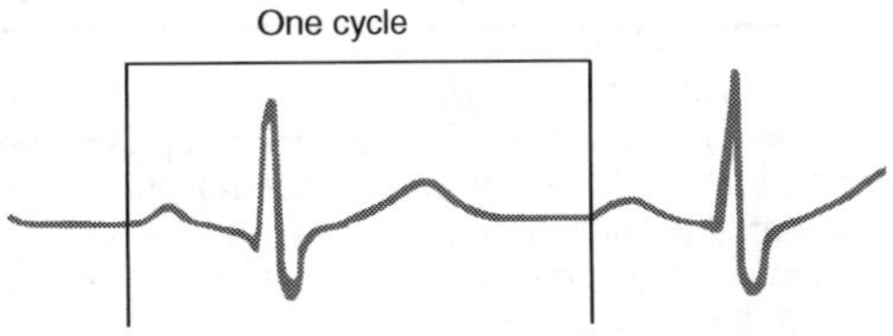

5. (a)

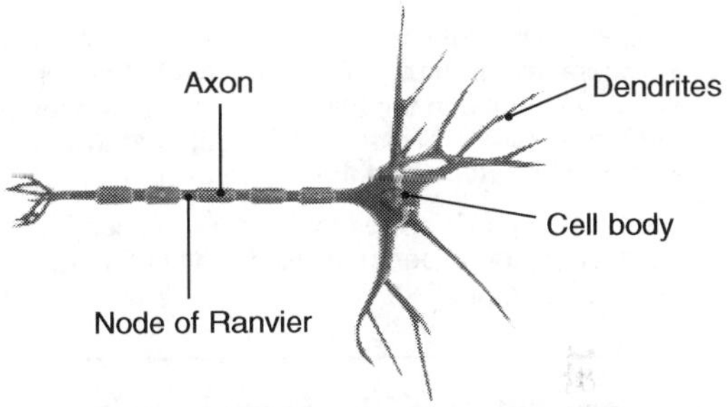

   (b) Myelinated
   (c) Schwann cells can be seen wrapped around the axon. Nodes of Ranvier can also be seen.
   (d) As action potentials (self-propagating depolarizations)

6. alveoli (I), breathing (D), diabetes (G), gas exchange (C), lungs (E), negative feedback (B), pancreas (H), respiratory gas (F), spirometry (A). Note the following clarifications:

   Alveoli definition should more correctly read "Microscopic structures in vertebrate lungs...." as the current wording implies some lungs do not have alveoli, which is not the case.

   Lungs definition should read "Internal gas exchange structures found in **most** vertebrates".

208. Packaging DNA in Nucleus (page 288)

1. (a) DNA: A long, complex nucleic acid molecule found in the chromosomes of nearly all organisms (some viruses have RNA instead). Provides the genetic instructions (genes) for the production of proteins and other gene products (e.g. RNAs).
   (b) Chromatin: Chromosomal material consisting of DNA, RNA, and histone and non-histone proteins. The term is used in reference to chromosomes in the non-condensed state.
   (c) Histone: Simple proteins that bind to DNA and help it to coil up during cell division. Histones are also involved in regulating DNA function.
   (d) Nucleosome: The basic unit of DNA packing consisting of a length of DNA wrapped around a bead of 8 histones.

2 (a) DNA methylation and histone modification can make DNA tightly pack together or remain loose depending on the type and placement of modification.
   (b) Tightly packed DNA cannot be transcribed while loosely packed DNA can be transcribed.

3. The chromatin is packaged up in an organized fashion. This occurs at several levels. The basic unit of DNA packing is the nucleosome, which is DNA wrapped around a bead of 8 histone proteins. These are separated by linker DNA sequences of about 50 bp. H1 histones pull nucleosomes together to form a chromatin fibre, which is then folded and wrapped so that it is held in a tight configuration. The different

ISBN: 978-1-927173-94-7

levels of coiling enables a huge amount of DNA to be packed into the nucleus, without tangling, in an orderly fashion..

209. DNA Molecules (page 290)

1. The gaps labelled A in the diagram made Watson and Crick realize that pattern was produced by a double helix.
2. 2%
3. (a) Protein coding regions are organized into introns (not translated) and exons (translated into proteins).
   (b) Introns may serve regulatory functions in the cell, e.g. as regulatory RNAs including ribosomal RNA (rRNA) and transfer RNAs (tRNA).

210. DNA Carries the Code (page 291)

1. Griffith first injected disease causing bacteria into healthy mice to confirm the action of the bacteria. The bacteria were then heated to kill them and injected into healthy mice. These mice did not develop pneumonia confirming it was the living bacteria that was causing the disease.
2. Sulfur is found in proteins but not in DNA (or to a much lesser extent) while phosphorus is found in DNA but not proteins. If the sulfur was found in the infected bacteria then the proteins carried the genetic information. If phosphorus was found then it was the DNA that carried the information.
3. Different experiments show that the result is not peculiar to, or the result of, the experiment itself. Rather it is a property of the system being studied.

211. Enzyme Control of DNA Replication (page 292)

1. Enzymes catalyse the reactions that occur during DNA replication. They unwind the DNA, copy the DNA strands, rejoin DNA sequences, and proofread the DNA to correct mistakes. Enzymes are important in ensuring the same sequence of DNA in the parent cell occurs in the daughter cells so that the cell functions normally.
2. (a) Helicase: Unwinds the ‘parental’ strands.
   (b) DNA polymerase I: Hydrolyses the RNA primer and replaces it with DNA.
   (c) DNA polymerase III: Elongates the leading strand. It synthesizes the new Okazaki fragment until it encounters the primer on the previous fragment.
   (d) Ligase: Joins Okazaki fragments into a continuous length of DNA.
3. $4 \times 10^6 \div 4000 = 1000$ s. $1000 \div 60 =$ **16 min 40 s**
4. The nucleotides are present as deoxynucleoside triphosphates. Hydrolysis provides the energy for incorporating them into the strand.
5. DNA polymerase can only work in the 5' to 3' direction so only the leading strand can be copied as a continuous piece. The lagging strand is copied in fragments, which are later joined.
6. (a) A dideoxyribonucleic acid
   (b) and (c)

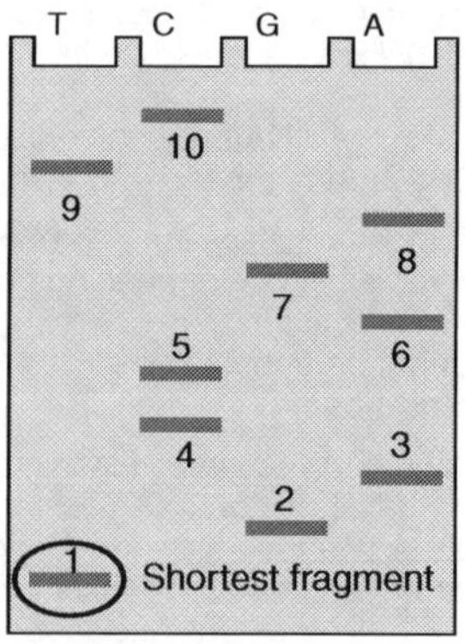

   (d) T G A C C A G A T C
   (e) A C T G G T C T A G

7. A reaction vessel is needed for each modified base. If all the modified bases were put in the same vessel there would be no way of distinguishing between them once placed on an electrophoresis gel.
8. 1% modified DNA is enough to produced terminated DNA fragments but is a small enough to allow other non-modified DNA to be incorporated in the DNA fragment. Too much modified DNA would cause termination of DNA replication too early in too many cases.

212. Transcription (page 294)

1. mRNA carries a copy of the genetic instructions from the DNA in the nucleus to ribosomes in the cytoplasm. The rate of protein synthesis can be increased by making many copies of identical mRNA from the same piece of DNA.
2. (a) Start codon: AUG
   (b) Stop codons: UAA, UAG, UGA
3. (a) AUG AUC GGC GCU AAA
   (b) AUG UUC GGA UAU UUU

213. Post Transcriptional Modification (page 295)

1. Primary mRNA can be spliced and combined in many different ways. This allows more proteins to be produced than the number of genes in the human genome. Post translational modification also allows for more protein variations to be produced.
2. There are four main ways in which mRNA can be modified to code for different proteins. (1) Exon skipping can leave out one or more parts of the mRNA. (2) Intron retention retains parts of the primary mRNA that would normally be left out. (3) Mutually exclusive exons can produce two alternative proteins, and (4) some exons have more than one binding site, which produces exons of different lengths and different proteins.
3. Being able to modify the mRNA allows an organism to produce alternative proteins without needing an entirely new gene to code for it. This ensures that the organism's genome is not too large. Note: For example, the Dscam gene in *D. melanogaster* can produce more than 38,000 proteins from the one gene.

214. Controlling Gene Expression (page 296)

1. (a) Transcription in prokaryotes is initiated when RNA polymerase and associated proteins bind to the promoter.
   (b) Transcription in eukaryotes is initiated when RNA polymerase is recruited by transcription factors and binds to the enhancer region of the DNA. Transcription factors bring the enhancer and promoter regions of DNA together to begin transcription.
2. The promoter is the point at which RNA polymerase binds to the DNA to begin transcription.

215. Gene-Environment Interactions (page 297)

1. Answer will vary based upon student's own choice, but examples include:

   Environment: The environment can alter the phenotype by altering the basic expression of a trait. Examples include: trees that are grown at high altitude are stunted in growth, trees exposed to a strong prevailing wind appear distorted in shape (windswept), poor nutrition in infancy can retard brain and bone development.

   Genotype: An organism's genotype has a major influence upon phenotype. Dominant alleles, recessive alleles, gene mutations, gene interactions, and epigenetics all influence phenotype. Examples include: albinism resulting from the inheritance of a recessive allele, the shape of the comb on a chicken depends on which allele combination they inherit, the colour of many flowers is determined by the dominance of one allele over another, and the colour of skin in humans is the result of six allele combinations.

ISBN: 978-1-927173-94-7

2. Wind speed, temperature, air density, water availability.
3. In hydrangeas, soil pH alters flower colour. In alkaline soils they are pink or red-purple, in acid soil the flowers are blue. The colour is due to the presence or absence of aluminium compounds in the flowers, and aluminium in the soil is only accessible to plants when soil pH is low. When aluminium is present within the plant, the flowers are blue. When the aluminium is absent, the flowers are pink. Note: Growers may apply aluminium sulfate to the soil to lower the pH and produce blue flowers. Adding lime produces pink flowers.
4. These are the cooler parts of the body. Body heat is lost from these areas which make it cool enough for the enzyme responsible for colour-pointing to remain active.
5. To ensure genetic potential is reached, provide the optimum growth conditions for that plant, e.g. suitable water availability (water regularly), adequate nutrient supply (fertilizer application), sufficient sunlight (keep out of shade), equable temperature (warmth, protect from frost and wind), protection from pests and diseases (companion plant or spray for pests).
6. (a) A cline is a continuous, or nearly continuous, gradation in a phenotypic character within a species, associated with a change in an environmental variable such as temperature or wind.
   (b) Plant A: The observed phenotype (prostrate) of this species is not due to genetic factors, but to the effect of climate on growth patterns. In the absence of a harsh environment, the plant reverts to its normal growing habit. Plant B: The low growing phenotype of this species is controlled by genes (not environmental factors). Species B continued to be low growing.
   (c) Plant A is most likely to show clinal variation.

216. DNA Methylation (page 299)

1. (a) DNA methylation is the adding of a methyl group ($-CH_3$) to a DNA base (usually cytosine).
   (b) Methylation of cytosine turns off gene expression, so that the gene is not transcribed.
2. The mother must have donated the imprinted (silenced) gene because if the father's gene was working correctly there would be no syndrome as there would still be a working copy of the gene. If we know the father's gene was mutated and the syndrome appears, then the mother's corresponding gene must be silenced.

217. Translation (page 300)

1. AUG AUC GGC GCU AAA
2. (a) 61
   (b) There are 64 possible codons for mRNA, but three are terminator codons. 61 codons for mRNA require 61 tRNAs each with a complementary anticodon.

218. Protein Synthesis Summary (page 301)

1. (a) 1: Unwinding the DNA molecule.
   (b) 2: mRNA synthesis: Nucleotides added to the growing strand of messenger RNA molecule.
   (c) 3: DNA rewinds into double helix structure.
   (d) 4: mRNA moves through nuclear pore in nuclear membrane to the cytoplasm.
   (e) 5: tRNA molecule brings in the correct amino acid to the ribosome.
   (f) 6: Anti-codon on the bottom of the tRNA matches with the correct codon on the mRNA and drops off the amino acid.
   (g) 7: tRNA leaves the ribosome.
   (h) 8: tRNA molecule is recharged with another amino acid, ready to participate in protein synthesis.
2. (a) DNA
   (b) Free nucleotides
   (c) RNA polymerase enzyme
   (d) mRNA
   (e) Nuclear membrane
   (f) Nuclear pore
   (g) tRNA
   (h) Amino acids
   (i) Ribosome
   (j) Polypeptide chain
3. (a) Whether or not the protein is required by the cell (regulated by control of gene expression).
   (b) Whether or not there is an adequate pool of the amino acids and tRNAs required for the protein in question.

219. Proteins (page 302)

1. (a) Peptide bonds between the amino acids.
   (b) α-helices and β-pleated sheets form due to hydrogen bonding.
   (c) Chemical bonds and hydrophobic interactions.
   (d) Interactions between two or more protein molecules or polypeptide chains (subunits).
2. The R groups on each amino acid allow weak intermolecular forces to bind different parts of the polypeptide chain together. This binding causes the polypeptide chain to fold up into the functional shape.

220. Chapter Review (page 303)

No model answer. Summary is the student's own.

221. KEY TERMS: Did You Get it? (page 304)

1. DNA replication (D), exon (C), helicase (H), intron (G), nucleosome (E), polysome (I), quaternary structure (B), secondary structure (A), transcription (J), translation (F)
2. DNA is a large molecule which forms a double helix. It stores the genetic code in units called genes. Each gene is composed of a number of codons. A codon is a sequence of three adjacent nucleotides. During transcription, the gene is copied into mRNA. The mRNA is read by ribosomes which match up the codons with their corresponding amino acids, binding them together to form proteins. The protein may be used structurally or it may act as an enzyme.
3. Changes in the environment cause genes to be switched on or off and so affect the development of the organism. Factors in the environment that can influence gene expression temperature, pH, and chemicals, as well as methylation of DNA due to environmental changes (epigenetic factors). For example the colour of the hydrangea flower is dependent on the pH of the soil. In many reptiles, the temperature at which the eggs are incubated affects the sex of the hatchlings.

222. Metabolic Pathways (page 306)

1. A metabolic pathway is a series of linked biochemical reactions which carry out processes essential to life.
2. Enzymes control each step of the metabolic pathway.
3. Fumarate, arginine, urea, ornithine

223. How Enzymes Work (page 307)

1. Enzymes are biological molecules (usually proteins) that act as catalysts, allowing reactions to proceed more readily. They do this by influencing bond stability in the reactants and thereby lowering the activation energy required to create an unstable transition state in the substrate from which the reaction proceeds readily.
2. The induced fit model is a modified version of the lock and key in which the substrate and the active site interact. Substrate binding causes the active site to change slightly so that bonds in the substrate(s) are destabilized. This model is supported by evidence from studies of enzyme inhibition.
3.

Mass (g)	Volume O_2 (cm^3)			Mean	Rate
	Test 1	Test 2	Test 3		
1	6	5	6	5.7	1.14
2	10	9	9	9.3	1.86
3	14	15	15	14.7	2.94
4	21	20	20	20.3	4.06
5	24	23	25	24	4.8

ISBN: 978-1-927173-94-7

4.

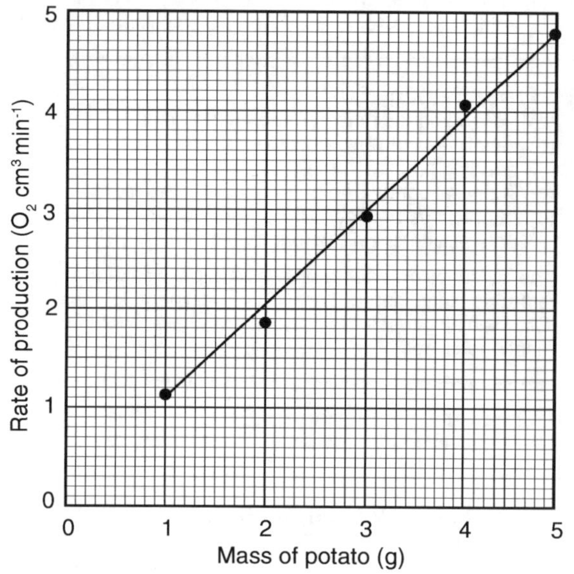

5. The rate of reaction increases with increasing enzyme present.
6. So that the H_2O_2 was not a limiting factor in the reaction.
7. The students should have used a control, that being a flask without potato added to the H_2O_2.
8. There would be no gas evolved.
9. Cooking the potato denatures the catalase enzyme.

224. Enzyme Inhibition (page 309)

1. In competitive inhibition, the inhibitor competes with the substrate for the enzyme's active site and, once in place, prevents substrate binding. A non-competitive inhibitor does not occupy the active site but binds to some other part of the enzyme, making it less able to perform its function as an effective biological catalyst.
2. Type of inhibition could be tested by increasing the substrate concentration. If this overcame the rate depression then the inhibition is competitive.
3. The MEP pathway is found in the malaria-causing parasite *P. falciparum* but not in humans. Therefore is can be targeted by drugs that will not affect human metabolic processes.

225. Control of Metabolic Pathways (page 310)

1. The end product of the metabolic pathway acts as an inhibitor for an enzyme that controls an early step in the pathway. As the concentration of end-product increases, the pathway becomes less active. In the threonine-isoleucine pathway, the end-product isoleucine inhibits the enzyme that converts threonine to alpha-ketobutyrate in the first step in the pathway. This regulates the pathway to prevent over-production of isoleucine. Please note our apologies for the typographical error in the spelling of threonine deaminase in the pathway.
2. The end-product of the metabolic pathway may act as the allosteric regulator which attaches to the allosteric site of an enzyme in the pathway and regulates its activity. When production of the regulator declines, it is released from the allosteric site and the pathway is reactivated.

226. The Biochemistry of Respiration (page 311)

1.

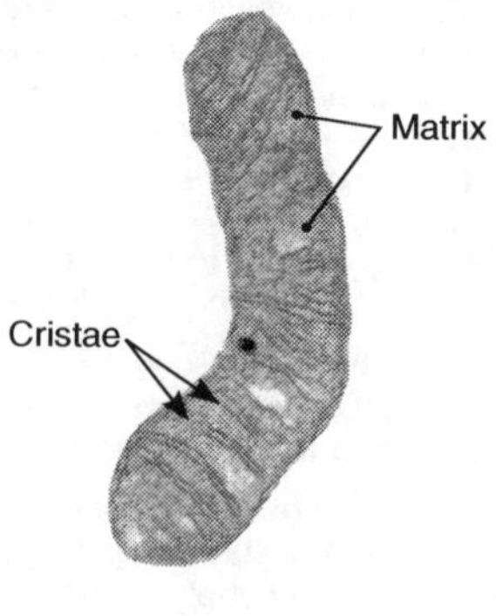

2. The link reaction prepares pyruvate for entry in the Kerbs cycle. Carbon dioxide is removed and coenzyme A is added.
3. (a) 6 (b) 3 (c) 2 (d) 6 (e) 5 (f) 4
4. (a) Glycolysis: 2 ATPs
   (b) Krebs cycle: 2 ATPs
   (c) Electron transport chain: 34 ATPs
   (d) Total produced: 38 ATPs
5. The carbon atoms lost are lost as CO_2 molecules.
6. During oxidative phosphorylation, ADP is phosphorylated to ATP. Electrons passed along the electron transport chain are used to pump hydrogen ions across the inner membrane of the mitochondria. The flow of hydrogen ions back across the membrane is coupled to the phosphorylation of ADP to ATP. Oxygen is the final electron acceptor, reducing hydrogen to water. Because oxygen is the final acceptor the process is called oxidative phosphorylation.

227. Chemiosmosis (page 313)

1. In chemiosmosis, ATP synthesis is coupled to electron transport and movement of hydrogen ions. Energy from the passage of electrons along the chain of electron carriers is used to pump protons (H^+), against their concentration gradient, into the intermembrane space, creating a high concentration of protons there. The protons return across the membrane down a concentration gradient via the enzyme complex, ATP synthase, which synthesizes the ATP.
2. Elevating the proton concentration outside the exposed inner mitochondrial membranes would them to move down their concentration gradient via ATP synthase and generating ATP.
3. A suspension of isolated chloroplasts would become alkaline because protons would be removed from the medium as ATP was generated.
4. (a) By placing chloroplasts in an acid medium, the thylakoid interior was acidified. Transfer to an alkaline medium established a proton gradient from the thylakoid interior to the medium.
   (b) The protons could flow down the concentration gradient established, via ATP synthase, and generate ATP.

228. Chloroplasts (page 314)

1. (a) Stroma (d) Granum
   (b) Stroma lamellae (e) Thylakoid
   (c) Outer membrane (f) Inner membrane
2. (a) Chlorophyll is found in the thylakoid membrane.
   (b) Chlorophyll is a component of the membrane-bound photosystems and captures light energy in the light dependent reactions of photosynthesis.
3. The internal membranes of chloroplasts provide a large surface area for binding chlorophyll molecules and capturing light. Membranes are stacked in such a way that they do not shade each other.
4. Chlorophyll absorbs blue and red light but reflects green light, so leaves look green to the human eye.

ISBN: 978-1-927173-94-7

229. Light Dependent Reactions (page 315)

1. NADP: Carries H_2 from the light dependent phase to the light independent reactions.
2. Chlorophyll: These pigment molecules trap light energy and produce high energy electrons. These are used to make ATP and NADPH. The chlorophyll molecules also split water, releasing H^+ for use in the light independent reactions and liberating free O_2.
3. Light dependent (D) phase takes place in the grana (thylakoid membranes) of the chloroplast and requires light energy to proceed. The light dependent phase generates ATP and reducing power in the form of NADPH. The electrons and hydrogen ions come from the splitting of water.
4. The ATP synthesis is coupled to electron transport. When the light strikes the chlorophyll molecules, high energy electrons are released by the chlorophyll molecules. The energy lost when the electrons are passed through a series of electron carriers is used to generate ATP from ADP and phosphate.

   Note: ATP is generated (in photosynthesis and cellular respiration) by chemiosmosis. As the electron carriers pick up the electrons, protons (H^+) pass into the space inside the thylakoid, creating a high concentration of protons there. The protons return across the thylakoid membrane down a concentration gradient via the enzyme complex, ATP synthase (ATP synthetase or ATPase), which synthesizes ATP.
5. (a) Non-cyclic (photo)phosphorylation: Generation of ATP using light energy during photosynthesis. The electrons lost during this process are replaced by the splitting of water.
   (b) The term non-cyclic photophosphorylation is also (commonly) used because it indicates that the energy for the phosphorylation is coming from light.
6. (a) In cyclic photophosphorylation, the electrons lost from photosystem II are replaced by those from photosystem I rather than from the splitting of water. ATP is generated in this process, but not NADPH. Note: In the cell, both cyclic and non-cyclic photophosphorylation operate to different degrees to balance production of NADPH and ATP.
   (b) The non-cyclic path produces ATP and NADH in roughly equal quantities but the Calvin cycle uses more ATP than NADPH. The cyclic pathway of electron flow makes up the difference.
7. It shows that a complex reaction pathway is made of less complex pathways that can operate independently. These simple pathways can then be linked through common intermediates, generating complex pathways.

230. Light Independent Reactions (page 317)

1. (a) 6 (b) 6 (c) 12
   (d) 12 (e) 12 (f) 6
   (g) 2 (h) 1
2. RuBisCo catalyses the reaction that splits CO_2 and joins it with ribulose 1,5-bisphosphate. It fixes carbon from the atmosphere.
3. Triose phosphate (note that you may also see this referred to as glyceraldehyde-3-phosphate, GALP, G3P or PGAL)
4. $6CO_2 + 18ATP + 12\ NADPH + 12H^+$
   $\rightarrow 1\ glucose + 18ADP + 18Pi + 12\ NADP^+ + 6H_2O$
5. The Calvin cycle will cease in the dark in most plants because the light dependent reactions stop, therefore no NADPH or ATP is produced. At night, stomata also close, reducing levels of CO_2 (there will still be some CO_2 in the leaf as a waste product of respiration).

231. Experimental Investigation of Photosynthesis (page 318)

1. $DCPIP + 2H^+ + 2e^- \rightarrow DCPIPH_2$
2. Hill's experiment showed that water must be the source of oxygen.
3. The reactions happen very quickly. By taking the sample only seconds apart each step of the reaction can be worked out by recording the order of appearance of the products of the reactions.

232. Chapter Review (page 319)

No model answer. Summary is the student's own.

233. KEY TERMS: Did You Get it? (page 320)

1. activation energy (F), Calvin cycle (D), chemiosmosis (A), competitive inhibition (B), electron transport chain (G), induced fit model (C), Krebs cycle (E), light dependent reactions (J), non-competitive inhibition (I), oxidative phosphorylation (H). Clarification: The definition for competitive inhibition should more correctly state..."A type of inhibition where the **substrate** and inhibitor compete....."
2. (a) True
   (b) False
   (c) True
   (d) False
3. (a) Glucose (e) ADP
   (b) Carbon dioxide (or water) (f) Amino acid
   (c) Water (or carbon dioxide) (g) Protein
   (d) ATP

234. The General Structure of Plants (page 323)

1. (a) **Stems**: Link the roots to the leaves. Provide support for the leaves, fruits, and flowers. Conduct water and dissolved minerals and foods around the plant. Produce new tissue at meristems.

   Leaves: Collect the sun's energy and convert it to usable energy (sugar). Control the entry and exit of gases and water vapor (therefore have a role in cooling the plant).

   Roots: Anchor the plant. Absorb water and dissolved minerals. Sometimes store food. Produce new tissue at meristems.
   (b) Materials transported: Water, minerals (e.g. N, P), sugar, essential ions (e.g. K^+, Na^+).
   (c) Functions of transport tissues: Xylem: Plant support and transport of water and dissolved minerals. Phloem: Plant support and transport of sugar (in solution), minerals, hormones, and amino acids.
2. Water
3. (a) Xylem (required answer in bold): **Osmosis** (passive movement of water across a partially permeable membrane from higher to lower concentration of water molecules). Also **tension-cohesion** and (to a lesser extent) root pressure.
   (b) Phloem (required answer in bold): **Active transport** to load the sugar into the phloem tissue, **osmosis** (water follows sugar into the phloem), **pressure-flow** of the sap along the phloem.

235. Uptake at the Root (page 324)

1. (a) Passive absorption of minerals along with the water and active transport.
   (b) Apoplastic pathway (about 90%); water moves through the xylem and the spaces within cell walls.
   Symplastic pathway; water moves through the cell cytoplasm from cell to cell via plasmodesmata.
2. Large water uptake allows plants to take up sufficient quantities of minerals from the soil. These are often in very low concentration in the soil and low water uptakes would not provide adequate quantities.
3. (a) The Casparian strip represents a waterproof barrier to water flow through the apoplastic pathway into the stele. It forces the water to move into the cells (i.e. move via the symplastic route).
   (b) This feature enables the plant to regulate ion uptake, i.e.

ISBN: 978-1-927173-94-7

absorb ions selectively. The movement of ions through the apoplast cannot be regulated because the flow does not occur across any partially permeable membranes.

236. Transpiration (page 325)

1. (a) The evaporative loss of water from the leaves and stem of a plant.
   (b) Any one of:
   – Transpiration stream enables plants to absorb sufficient quantities of the minerals they need (the minerals are absorbed with the water and are often in low concentration in the soil).
   – Transpiration helps cool the plant.
2. Water loss is regulated by the opening and closing of stomata.
3. (a) The plant would lose water from the cells and wilt.
   (b) During a prolonged period without water (e.g. a drought).
4. Water moves by osmosis in all cases. In any order:
   (a) Transpiration pull: Photosynthesis and evaporative loss of water from leaf surfaces create higher solute concentrations (lower water concentration) in the leaf cells than elsewhere, facilitating movement of water down its concentration gradient towards the site of evaporation (stomata).
   (b) Cohesion-tension: Water molecules cling together and adhere to the xylem, creating an unbroken water column through the plant. The upward pull on the water creates a tension that facilitates movement of water up the plant.
   (c) Root pressure provides a weak push effect for upward water movement.
5. Water is moved up the tree by a combination of cohesion-tension, transpiration pull, and root pressure, Together these process can move water up to heights of far more than 40 m.

237. Xylem (page 327)

1. (a) Xylem conducts water and dissolved minerals around the plant (from roots to leaves).
   (b) Xylem forms a continuous tube for the passive process of water transport. Because the transport is passive the cell does not need to be alive.
2. (a) Vessel elements: Provide rapid, low resistance conduction of water.
   (b) Tracheids: Role in water conduction but there is more impedance to water flow than in vessels.
   (c) Fibres (and sclereids): Provide mechanical support to the xylem.
   (d) Xylem parenchyma: Involved in storage.
3. Vascular bundle showing xylem. Other regions are shown in plan view only.

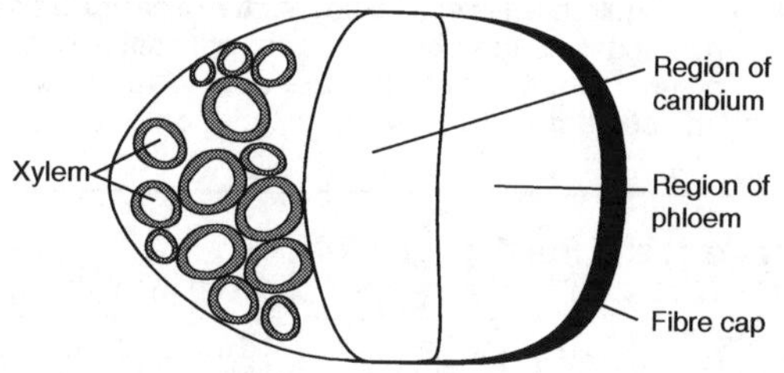

238. Investigating Plant Transpiration (page 328)

1. (a) Plot of data below.

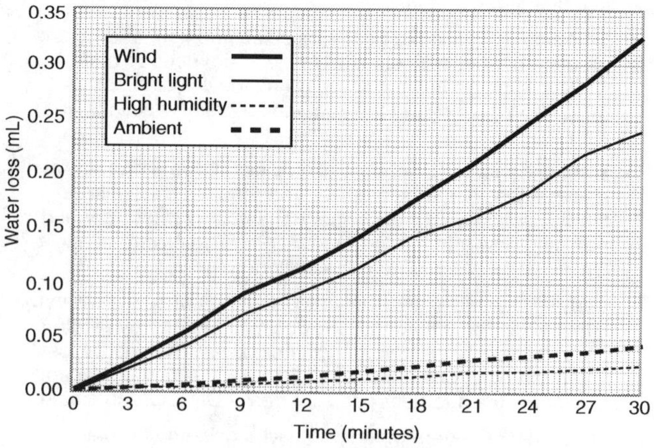

   (b) Independent variable: Time. Explanatory note: Environmental conditions are manipulated in that there are different treatments, but each condition constitutes a controlled variable (or treatment).
2. (a) Transpiration rate in ambient conditions.
   (b) An experimental control enables a measure of the biological response in the absence of any of the manipulated variables being tested (no treatment). This serves as a reference point.
   (c) Wind and bright light increased water loss above the ambient (control) conditions.
   (d) Wind and bright light increase transpiration rate by increasing evaporative loss from the leaves. High humidity reduces transpiration loss by reducing the gradient in water concentration from leaf to air (this decreases evaporation).
   (e) In humid conditions, there is a reduced gradient in free water concentration between leaf and air, so the rate of diffusion of water vapour from the leaf to the environment will be slower.

239. Adaptations for Water Conservation (page 330)

1. Xeromorphic adaptations allow xerophytes to survive and grow in areas with low or irregular water supplies.
2. (a)-(c) any of:
   – Modification of leaves to reduce transpirational loss (e.g. spines, curling, leaf hairs).
   – Shallow, but extensive fibrous root system to extend area from which water is taken and to take advantage of overnight condensation.
   – Water storage in stems or leaves.
   – Rounded, squat shape of plant body to reduce surface area for water loss.
3. The CAM metabolism (found only in xerophytic plants, many of which are succulents) allows CO_2 to be fixed during the dark. This produces organic acids which accumulate in the leaves and later release CO_2 into the Calvin cycle during daylight (when light energy is available to provide H+ and ATP for photosynthesis). The stomata can then remain closed during the day when water losses are highest.
4. A moist microenvironment reduces the gradient in water potential between the leaf and the air, so there is less tendency for water to leave the plant.
5. In a high salt environment (often also exposed and windy), free (physiologically available) water is in short supply. Sea shoreline plants therefore have many xeromorphic adaptations.

240. Translocation (page 332)

1. (a) The increase in dissolved sugar in the sieve tube cell decreases its free water concentration. Water then moves into the sieve tube cells by osmosis, creating a pressure that pushes the sugar solution through the phloem.

ISBN: 978-1-927173-94-7

(b) This means the sugar flows from its site of production (leaves) to its site of unloading (roots).

2. Food is manufactured in one region of the plant (the leaves) but is required in other regions (e.g. the roots and fruits). It must be transported there.

3. (a) Translocation: The transport (around the plant) of the organic products of photosynthesis.
   (b) The bulk movement of phloem sap along a gradient in hydrostatic pressure (generated osmotically).
   (c) The coupling of sucrose transport (into the transfer cell) to the diffusion of H^+ down a concentration gradient (generated by a proton pump).

4. Sucrose is transported into the phloem by a protein symport. Sucrose is coupled to the diffusion of hydrogen ions into the cell.

5. The transfer cell uses active transport mechanisms (coupled transport of sucrose) to accumulate sucrose to levels 2-3 times those in the mesophyll. The sucrose then moves into the sieve tube cell.

6. If sap moved by pressure-flow, then there should be selective pressure for the sieve plate to be lost or become less of a barrier, yet this has not happened. Of course, there are also selective pressures that operate against loss of the sieve plate, such as the need to have discrete yet freely communicating cells.

241. Phloem (page 334)

1. Phloem conducts dissolved sugar around the plant from its place of production to where it is required.
2. The phloem needs transport proteins to actively move dissolved nutrients such as sucrose into the phloem stream, The cells must therefore be alive with active plasma membranes to achieve this. Movement of minerals into the xylem in passive and requires no transport proteins.
3. The companion cell keeps the sieve tube cell alive and controls its activity. They are responsible for loading and unloading sugar into the sieve cells.

242. Identifying Xylem and Phloem (page 335)

1.

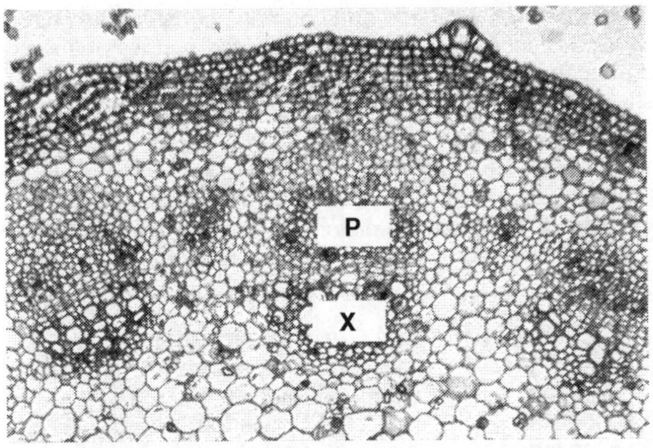

2.

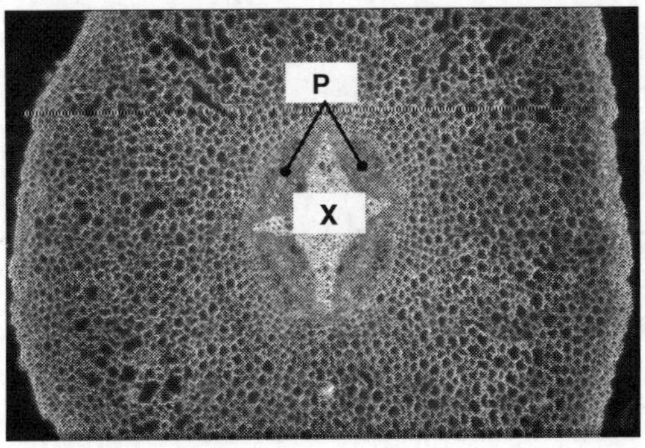

3. A: Fibre cap
   B: Phloem
   C: Xylem
   D: Cortex
   E: Vascular cambium
   F: Pith

243. Plant Meristems (page 336)

1. The meristems in plants consist of undifferentiated cells and cell division in the meristems provide new cells for expansion and differentiation of tissues.
2. Primary meristems are located in apical regions, i.e. the growing tips of the shoots and roots. Primary meristems are responsible for an increase in length or height. Some plants also have secondary meristems (the vascular cambium) encircling the stem, which produces increase in plant girth (diameter).
3. Meristematic tissue comprises undifferentiated cells (it is equivalent to stem cells in animals).
4. Primary tissues are formed from the primary meristems. Three primary meristems are produced from the apical meristem. The protoderm forms the epidermis, the ground meristem forms pith, cortex, and mesophyll, while the procambium leads to the formation of primary phloem and xylem. These form vascular bundles.

244. Auxins and Shoot Growth (page 337)

1. Positive phototropism
2. In an experiment in which a cut stem with an auxin infused agar block is uppermost and an agar block without auxin is at the base, auxin moves down the stem. However, if the system is inverted, no auxin is found in the stem, indicating that the auxin in the agar was not transported or diffused through the stem - it only travels one way.
3. Auxin causes cell elongation.

245. Micropropagation of Plant Tissue (page 338)

1. The purpose of micropropagation is to produce large numbers of clones (with identical phenotypic and genetic traits as the parent) in a short space of time.
2. (a) A callus is a mass of undifferentiated cells.
   (b) Several plant hormones are added to the culture in sequence. These will stimulate each phase of plant development.
3. Continued culture of a limited number of cloned varieties leads to a change in the genetic composition of the population (the amount of genetic variation decreases). This reduces the amount of variation upon which the gene pool can draw in times of change. This in turn reduces the ability (of the gene pool) to adapt.
4. Tissue culture has a number of advantages over traditional propagation methods. Tissue culture enables the production of many clones from a single seed/explant and allows the selection of desirable traits directly from culture. It facilitates rapid propagation, with no wait for seed production and is ideal for plants with long generation times, low seed production, or seeds that are difficult to germinate. Tissue culture also facilitates the international exchange of plants without quarantine and allows researchers to eliminate plant diseases from propagation lines. It is also space saving and overcomes seasonal restrictions to propagation. However, there are several disadvantages to this technology. Tissue culture is very labour intensive and trial and error is necessary to determine the ideal culturing conditions. Cultured plants may be genetically unstable/infertile and, over time, there may be a loss of genetic diversity.

246. Insect Pollinated Flowers (page 340)

1. The stigma is the receptive part of the female reproductive unit (the carpel), whereas the anther is the male organ of reproduction (produces the pollen, the male gametophyte).
2. Flowers attract pollinators by using bright colour, scent and

ISBN: 978-1-927173-94-7

energy (food) incentives such as nectar.

3. The plant must pass from the juvenile to sexually mature adult, the apical meristem must change from a vegetative meristem to a floral meristem, and the flowering organs must grow and develop.

247. Flowering (page 341)

1. (a) Day length (perhaps more importantly, night length).
   (b) Any one of:
   – Flowering at the same time ensures that other flowers will be available to provide/receive pollen.
   – Synchronization with periods of high insect activity may assist pollination.
2. (a) Short-day plants: Flower only when the day length is short (average: 10 hours).
   (b) Long-day plants: Flower only when the day length exceeds a certain minimum value (average: 14 h).
3. Short-day plants are really long night plants, requiring a night length of more than a minimum value. In the experiment outlined, a short-day plant failed to flower when a long night was interrupted by a short period of light (the plant interpreted this as a short night irrespective of the short day prior to it).

248. Control of Flowering (page 342)

1. (a) Pr (660nm) and Pfr (730nm)
   (b) Pfr is the active form. In long day plants it promotes flowering. In short day plants it inhibits flowering.
2 (a) Pr converts rapidly to Pfr in light. Pfr converts more slowly to Pr in darkness. The relative concentrations of Pr and Pfr give the plant the ability to measure day length. The phytochromes also interact with clock genes which maintain the plant's biological clock.
   (b) The ability to measure time gives all the plants of a particular species the ability to determine the appropriate time for flowering. All plants will behave the same way and thus flower at the same time.

249. Pollination Relationships (page 343)

1. A relationship in which both parties benefit.
2. The pollinator gains energy from the nectar or pollen eaten, while the flower benefits from having its pollen transferred directly to the stigma of another flower.
3. (a) Plants use scent, nectar, and bright colours to attract pollinators. Some flowers have evolved to imitate the female parts of an insect species, attracting the male insect, which picks up pollen while attempting to mate with the flower.
   (b) Advantage: There are many different pollinators available to transfer pollen.
   Disadvantage: A generalist pollinator may not transfer pollen from one plant to another of the same species, so pollen may be wasted.

250. Pollination and Fertilization (page 344)

1. Pollination refers specifically to the transfer of pollen from the male anthers to the female stigma. In contrast, fertilization in plants refers to the fusion of one of the sperm nuclei with the egg to form the embryo.
   Note: At the same time, another sperm nucleus fuses with the polar nuclei (female) to produce the endosperm
2. Plants increase the chances of pollination by (1) attracting pollinators (often specific) to move pollen from the anther to the stigma of another plant and (2) producing vast amounts of pollen, as in wind pollinated plants.
3. Pollen germinates on the stigma and grows a pollen tube towards the ovary, which enters the ovule through the micropyle. The sperm nuclei travel down the pollen tube. One fertilizes the egg cell the other carries out a double fertilization of the polar nuclei in the endosperm.

251. Seed Dispersal (page 345)

1. (a) Mechanism: Wind.
   (b) Adaptive features: Very light seeds with feathery heads to capture the air currents.
2. (a) Mechanism: Germinate where they fall. As they are heavy they do not move far from the parent plant. They are much sought after by animals as food.
   (b) Adaptive features: High energy, readily accessible food for many animals. Only the acorns surviving this will germinate. The acorn is heavy and falls to the ground. Because so many are eaten, those that escape predation have a better chance of vigorous growth. Rodents will also carry acorns to stores and then lose track of where they are. These will germinate if not too deeply buried.
3. (a) Mechanism: Water.
   (b) Adaptive features: The fruits have thick fibrous walls containing air pockets, enabling them to float. The seed inside remains unaffected by the salt water and germinates once it is washed ashore.
4. (a) Mechanism: Wind
   (b) Adaptive features: Light seeds attached to "wings" which act like a propeller in the air currents, carrying the seed away from the parent plant.
5. (a) Mechanism: Animal.
   (b) Adaptive features: The fleshy structures surrounding the seeds are eaten by insects such as ants, and the seeds are then abandoned.
6. (a) Mechanism: Wind.
   (b) Adaptive features: Pods split and shed many light, flaky seeds easily blown about by the wind.

252. A Most Accomplished Traveller (page 346)

1. Determining the origin of the coconut palm is difficult because it is so widespread and has been moved by humans as well as by floating across oceans.
2. Features include any of: The endosperm takes up little space inside the seed. An air space inside the seed adds to buoyancy. The seed is never dormant but there is a long germination time appropriate to long periods of sea travel. The seed is surrounded by a fibrous husk that keeps out seawater. The oval shape of the fruit allows it to ride high in the water. The seed is large, but has a thin, tough shell.
3. Coconuts were useful to early humans because they were easily transportable and provided a food and water source protected inside a shell that would keep for many days. The shell and fibrous outer layer also had many uses as bedding and for ropes.
4. The Atlantic Ocean is not connected to the Pacific or Indian Oceans except at the polar regions. The coconut did not reach the Atlantic because the distances it would have had to travel are too great and the seed would not have remained viable.

253. Seed Structure and Germination (page 347)

1. (a) Seed coat (f) Endosperm
   (b) Plumule (g) Plumule
   (c) Radicle (h) Cotyledon
   (d) Cotyledon (i) Radicle
   (e) Seed coat (j) Root cap

Error in workbook (first printing only): Pointers for monocot seed have been displaced from their correct positions in the first printing of this workbook. The corrected labels are shown right.

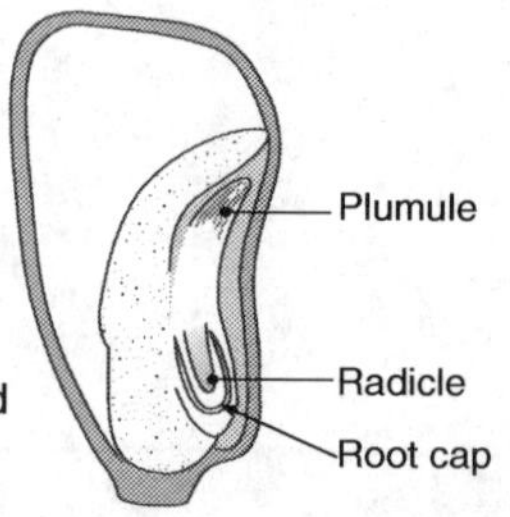

2. A seed is a reproductive unit. It houses the dormant embryonic plant surrounded in a protective seed coat.

ISBN: 978-1-927173-94-7

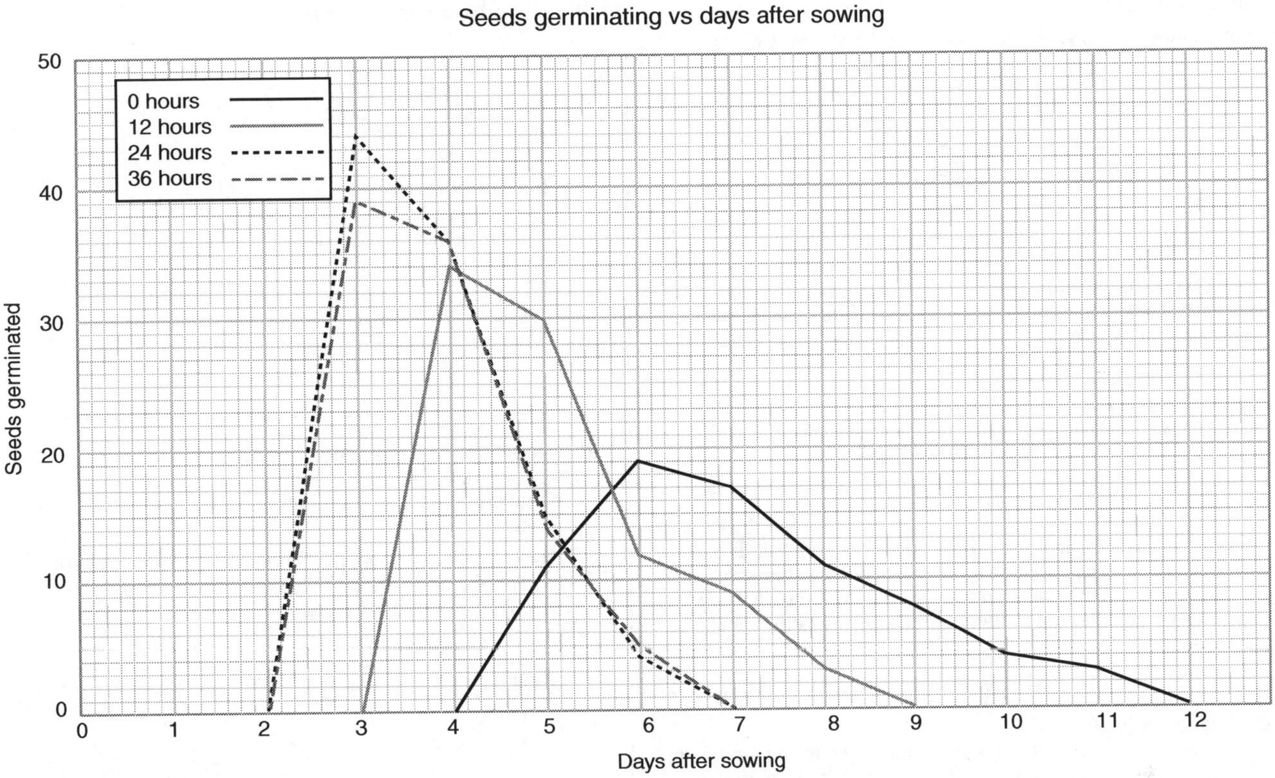

3. (a) The endosperm is rich in nutrients and provides food for the developing embryo. Note: In most monocots, it also provides nutrients to the seedling. In dicots, the endosperm nutrients are restocked into the cotyledons, which provide the food store.
   (b) The (triploid) endosperm is derived from the fertilization of the diploid endosperm nucleus.
4. Testa forms a protective covering around the seed.
5. If stored seeds become wet they begin to respire and germinate. Teacher's note: In storage, the heat generated by respiration denatures the seed proteins and germination stops again. The seeds are spoiled because their food store is gone.

254. Investigating Germination (page 348)

1. See graph at the top of the page.
2. (a) 24 hours (b) 24 hours
3. The experiment could have been carried out with more replicates per soaking period (e.g. three sets of seeds for each soaking time).
4. Experiments could include: Changing the temperature of the water or investigating the effect on other species or varieties.

255. Chapter Review (page 349)

No model answer. Summary is the student's own.

256. KEY TERMS: Did You Get it? (page 350)

1. auxin (E), cohesion-tension (C), flower (J), long-day plant (L), meristem (I), phloem (G), phytochrome (H), pollination (A), potometer (D), short-day plant (B), stomata (K), xylem (F)

   Clarification: definition for short day plant should more correctly read: " Plant that flowers in response to a period of dark exceeding a certain length."
2. (a) Sieve tube end plate
   (b) Phloem
   (c) Alive
   (d) Translocation or active transport of sugar.
   (e) Sugars or plant sap
3. (a) Positive phototropism
   (b) Directional light
   (c) Auxin
4. (a) Dicotyledons
   (b) Monocotyledons

257. Meiosis and Variation (page 352)

1. Independent assortment refers to the random distribution of maternal and paternal homologues to the gametes. This results in 2^n possible combinations of maternal and paternal chromosomes in gametes, where n is the haploid number, i.e. a large number of possible combinations.
2. (a) Crossing over is the mutual exchange of pieces of chromosome (alleles) between non-sister chromatids of homologous chromosome pairs.
   (b) Crossing over increases variation by creating new combinations of alleles on the chromosomes involved in the crossing over. The more crossing over incidents, the greater the variation.

258. Crossing Over (page 353)

1. (a) Unexpected combinations of alleles for genes will be present in gametes (recombinants).
   (b) The offspring are more genetically variable than they would otherwise be.
2. Crossing over provides genetic variation amongst individuals and this variation provides the raw material on which natural selection acts.

259. Crossing Over Problems (page 354)

Each problem stands alone (they are not a sequence).

1. (a) Gene sequences after crossing over at point 2:

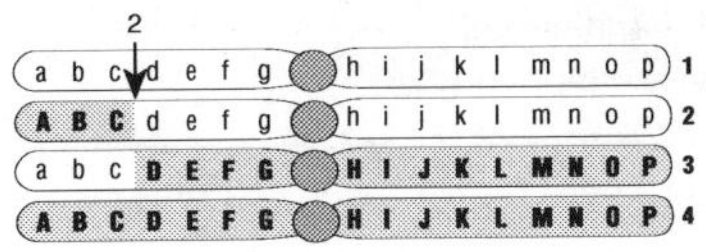

ISBN: 978-1-927173-94-7

(b) A, B and C

2. (a) Gene sequences after crossing over at points 6 & 7:

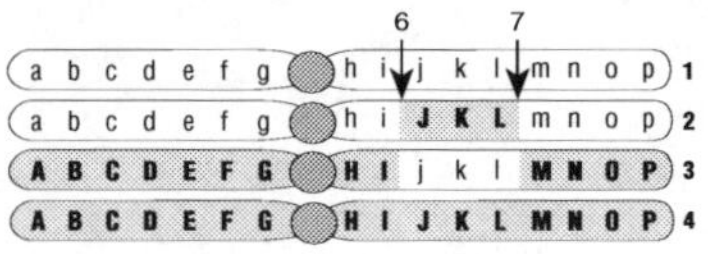

(b) J, K and L

3. (a) Gene sequences after crossing over at points 1 and 3, 5 and 7. (Note that results for chromatids 2 & 3 are interchangeable):

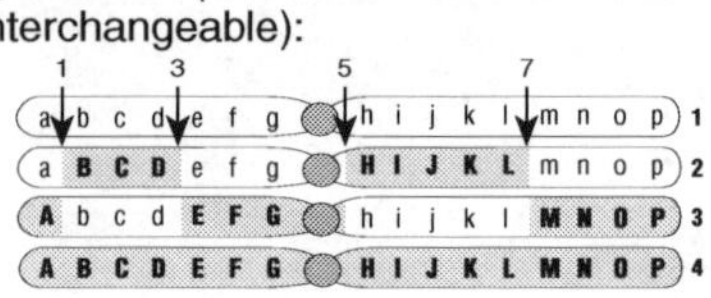

(b) B, C, D and H, I, J, K, L

4. Genes on one chromosome would be inherited together and there would be less variation in the gametes and therefore in the offspring. Variation would arise only from mutation and from the combination of maternal and paternal gametes at fertilization.

260. Variation (page 355)

1. The environment can alter the phenotype by altering the basic expression of a trait. Examples include: trees that are grown at high altitude are stunted in growth, trees exposed to a strong prevailing wind appear distorted in shape (windswept), poor nutrition in infancy can retard brain and bone development.

2. Variation is the raw material on which natural selection acts. Selection acts on the phenotype (being the result of genotype and environment). Selection acts for or against particular allele combinations, enhancing or reducing the representation of these in the next generation. Selection therefore sorts the variability in a gene pool and establishes adaptive genotypes.

3. (a) Meiosis provides an assortment of different gametes as a result of independent assortment.
   (b) Crossing over in meiosis exchanges parts of chromosomes and results in genes being swapped on chromosome pairs.
   (c) Fertilization will bring together the genes of two different individuals.

4. (a) It is one that has no harmful or beneficial effect under the current conditions.
   (b) They can be carried without selection pressure and may be beneficial or harmful under a different set of environmental conditions in the future.

5. Continuous variation is characterized by a very large number of phenotypic variations (so that a large sample of the population would exhibit a normal distribution for the trait in question). Discontinuous variation is characterized by a limited number of phenotypic variants. Teacher's note: Traits exhibiting continuous variation are determined by a large number of genes and are also frequently influenced by environment, e.g. hand span, weight, skin colour, height. Traits exhibiting discontinuous variation are determined by a single gene and include features such as chin dimple (present/absent) and blood groups (A, B, AB, O).

6. (a) Wool production: Continuous
   (b) Hand Span: Continuous
   (c) Blood groups: Discontinuous
   (d) Albinism: Discontinuous
   (e) Body weight: Continuous
   (f) Flower colour: Discontinuous

261. Dihybrid Cross (page 357)

1

	BL	Bl	bL	bl
BL	BBLL	BBLl	BbLL	BbLl
Bl	BBLl	BBll	BbLl	Bbll
bL	BbLL	BbLl	bbLL	bbLl
bl	BbLl	Bbll	bbLl	bbll

1 BBLL 2 BbLL 2 BBLl 4 BbLl
1 BBll 2 Bbll
1 bbLL 2 bbLl
1 bbll

262. Inheritance of Linked Genes (page 358)

1. Linkage refers to the situation where genes are located on the same chromosome. As a result, the genes tend to be inherited together as a unit.

2. Gene linkage reduces the amount of variation because the linked genes are inherited together and fewer genetic combinations of their alleles are possible.

3.

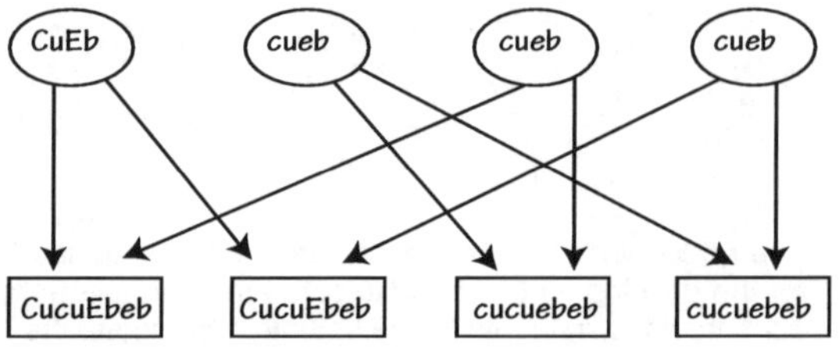

4. (a) CucuEbeb, Cucuebeb, cucuEbeb, cucuebeb
   (b) Offspring genotype: All CucuEbeb (heterozygotes)
   Offpspring phenotype: All wild type (straight wing, grey body)

5. Female gametes = VgEb and vgEb
   Male gametes = vgeb
   Offspring:
   Genotypes: VgvgEbeb vgvgEbeb
   Phenotypes: straight wing, grey body. Vestigial wing, grey body

6. *Drosophila* produce a wide range of mutations, have a short reproductive cycle, produce large numbers of offspring and are easy to maintain.

263. Recombination and Dihybrid Inheritance (page 360)

1. It produces new associations of alleles in offspring.

2.

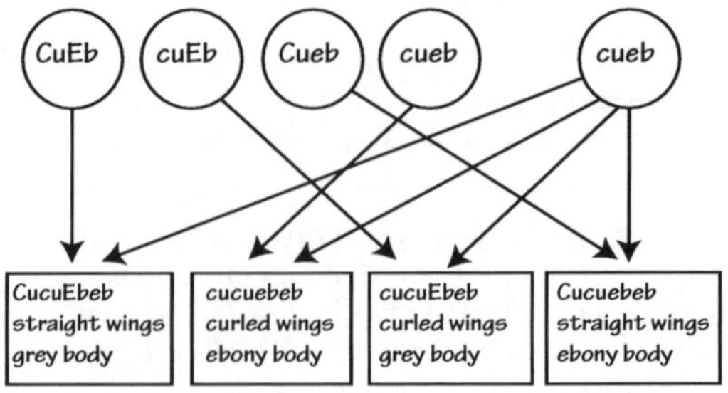

3. Parental linkage groups separate and new associations of alleles are formed in the offspring. The offspring show new combinations of characters that are unlike the parental types.

4. A greater than 50% recombination frequency indicates that there is independent assortment (the genes must be on separate chromosomes).

ISBN: 978-1-927173-94-7

5. Female gametes: CuY , cuY
Male gametes: Cuy, cuy
Offspring genotypes and phenotypes:
CuCuYy, CucuYy, CucuYy: Straight wings grey body
cucuYy: Curly wings grey body

264. Detecting Linkage in Dihybrid Inheritance (page 362)

1. Expected ratios as follows:

Purple, long (P_L_)	215
Purple, round (P_ll)	71
Red, long (ppL_)	71
Red, round (ppll)	24
Total 381	

2. (a) Expected ratios all 710 for each genotype as a 1:1:1:1 ratio is expected.
(b) Parental (given), recombinant, recombinant, parental.
(c) Morgan performed a test cross.

3. (a) Nail-patella syndrome is dominant. We can tell this because nearly all of the affected individuals had at least one parent with the disease.
(b) The affected parent was blood group B. All of their offspring with the B blood group also had nail-patella syndrome. Therefore, nail-patella syndrome is linked to the B blood group allele.
(c) Individual III-3 does not have nail-patella syndrome despite having the B blood type. It is likely that recombination has occurred, so this individual has not received the nail-patella gene.

265. Problems Involving Dihybrid Inheritance (page 363)

1. (a)

	BL	Bl
Bl	BBLl	BBll
bl	BbLl	Bbll

Genotype ratio: 1BBLl: 1BBll: 1BbLl: 1Bbll
Phenotype ratio: 1 black short hair, 1: black long hair

(b)

	TL	Tl	tbL	tbl
tbL	TtbLL	TtbLl	tbtbLL	tbtbLl
tbl	TtbLl	Ttbll	tbtbLl	tbtbll

Genotype ratio:

1TtbLL : 2TtbLl : 1Ttbll : 1tbtbLL : 2tbtbLl : 1tbtbll
Phenotype ratio: 3: Tabby long hair, 1:Tabby short hair, 3: blotched tabby long hair, 1: blotched tabby short hair

2. (a) Self pollination of plants with orange striped flowers produces progeny in ratios close to 9:3:3:1 (the expected ratio of a cross between heterozygous offspring of true breeding parents). Thus you may hypothesize that O (orange petals) is dominant to o (yellow petals) and stripes (S) are dominant to no stripes (s).
(b) The plants with the orange striped flowers were genotype OoSs. You know this because OoSs x OoSs will produce the progeny phenotypes in the observed 9:3:3:1 ratio.

3. (a) bbSS (brown/spotted) X BBss (solid/black) (which parent was male and which female is unknown. Parents must be homozygous since all the offspring are of one type: BbSs: black spotted).
(b) F2 generation: BbSs X BbSs

	BS	Bs	bS	bs
BS	BBSS	BBSs	BbSS	BbSs
Bs	BBSs	BBss	BbSs	Bbss
bS	BbSS	BbSs	bbSS	bbSs
bs	BbSs	Bbss	bbSs	bbss

(c) Spotted/black 9/16
Spotted/brown 3/16
Solid/black 3/16
Solid/brown 1/16
Ratio: 9:3:3:1 (described as above)

4. (a) F1: Genotype: all heterozygotes RrBb.
(b) F1: Phenotype: all rough black coats.
(c) F2 generation: RrBb X RrBb

	RB	Rb	rB	rb
RB	RRBB	RRBb	RrBB	RrBb
Rb	RRBb	RRbb	RrBb	Rrbb
rB	RrBB	RrBb	rrBB	rrBb
rb	RrBb	Rrbb	rrBb	rrbb

(d) Rough/black 9/16 Smooth/black 3/16
Rough/white 3/16 Smooth/white 1/16
Ratio: 9:3:3:1 (described as above)
(e)

	RB	Rb	rB	rb
RB	RRBB	RRBb	RrBB	RrBb

(f) F2 Phenotype: all rough black coats.
(g) The parents' genotypes: RrBb X Rrbb

5. Note: Persian and Siamese parents are pedigrees (truebreeding) and homozygous for the genes involved.
(a) Persian: UUss, Siamese: uuSS, Himalayan: uuss
(b) F1: Genotype: all heterozygotes UuSs.
(c) F1: Phenotype: all uniform colour, short haired.
(d) F2 generation: UuSs X UuSs

	US	Us	uS	us
US	UUSS	UUSs	UuSS	UuSs
Us	UUSs	UUss	UuSs	Uuss
uS	UuSS	UuSs	uuSS	uuSs
us	UuSs	Uuss	uuSs	uuss

(e) 1:15 or 1/16 uuss: Himalayan
(f) Yes (only one type of allele combination is possible)
(g) 3:13 or 3/16 (2 uuSs, 1 uuSS)
(h) All of the following have different genotypes but produce a uniform colour-short hair cat:

UUSS, UuSS, UuSs, UUSs, because they all have at least one dominant allele for each gene. Similarly uuSs and uuSS both produce a colour pointed short hair cat, and UUss and Uuss both produce a uniform coloured, long hair cat.

6. (a) Yes
(b) Four phenotypes were produced. If there was no crossing over there would only be two phenotypes (parental types).
(c) CucuEbeb, cucuebeb, Cucuebeb, cucuEbeb.

267. Chi-Squared Exercise in Genetics (page 366)

1. (a) H0: "If both parents are heterozygous and there is independent assortment of alleles (no linkage) then we would expect to see a 9:3:3:1 ratio of phenotypes in the offspring ".
(b) HA: "If both parents are heterozygous and the genes are linked (i.e. on the same chromosome), then we would expect the ratios of phenotypes in the offspring to deviate from the 9:3:3:1".

ISBN: 978-1-927173-94-7

2. (a) Completed table:

Category	O	E	O – E	$(O-E)^2$	$\frac{(O-E)^2}{E}$
Purple stem, jagged leaf	12	16.3	–4.3	18.5	1.1
Purple stem, smooth leaf	9	5.4	3.6	13	2.4
Green stem, jagged leaf	8	5.4	2.6	6.8	1.3
Green stem, smooth leaf	0	1.8	–1.8	3.2	1.8
	Σ 29				Σ 6.6

Expected frequencies calculated as follows:
Purple stem, jagged leaf = 9/16 x 29 = 16.3
Purple stem, smooth leaf = 3/16 x 29 = 5.4
Green stem, jagged leaf = 3/16 x 29 = 5.4
Green stem, smooth leaf = 1/16 x 29 = 1.8
Note: Whole numbers could be used in preference to rounding to one decimal place.

(b) $\chi^2 = 6.6$

(c) Degrees of freedom = (4-1=) 3

(d) The critical value of χ^2 at P = 0.05 and at d.f.= 3 is 7.82. The calculated χ^2 value is less than the critical value (6.6 < 7.82).

(e) We cannot reject H_0: There was no significant difference between the observed results and the expected phenotype ratio of 9:3:3:1. We must conclude that the genes controlling stem colour and leaf shape in tomatoes are on separate chromosomes (unlinked).

3. (a) H_0 and H_A as for question 1.

(b) Completed table:

Category	O	E	O – E	$(O-E)^2$	$\frac{(O-E)^2}{E}$
Round-yellow seed	441	450	–9	81	0.18
Round-green seed	159	150	9	81	0.54
Wrinkled-yellow seed	143	150	–7	49	0.33
Wrinkled-green seed	57	50	7	49	0.98
	Σ 800				Σ 2.03

Expected frequencies calculated as follows:
Round-yellow seed = 9/16 x 800 = 450
Round-green seed = 3/16 x 800 = 150
Wrinkled-yellow seed = 3/16 x 800 = 150
Wrinkled-green seed = 1/16 x 800 = 50
$\chi^2 = 2.03$

(c) Degrees of freedom = (4-1) 3 . The critical value of χ^2 at P = 0.05 and at d.f.= 3 is 7.82. The calculated χ^2 is less than the critical value (2.03 < 7.82).

(d) We cannot reject H_0: There was no significant difference between the observed results and the expected phenotype ratio of 9:3:3:1. We must conclude that the genes controlling seed shape and colour are on separate chromosomes (unlinked).

4. In both cases, we cannot reject H_0, but in the first case, the χ^2 value is much higher. In the case of the tomatoes, the genes for stem colour and leaf shape are on separate chromosomes, but given the relatively large χ^2 value, repeating the experiment using more plants, or replicate setups, would serve as a check.

268. Polygenes (page 367)

1. Punnett square

Gametes	ABC	ABc	AbC	Abc	aBC	aBc	abC	abc
ABC	AABBCC	AABBCc	AABbCC	AABbCc	AaBBCC	AaBBCc	AaBbCC	AaBbCc
ABc	AABBCc	AABBcc	AABbCc	AABbcc	AaBBCc	AaBBcc	AaBbCc	AaBbcc
AbC	AABbCC	AABbCc	AAbbCC	AAbbCc	AaBbCC	AaBbCc	AabbCC	AabbCc
Abc	AABbCc	AABbcc	AAbbCc	AAbbcc	AaBbCc	AaBbcc	AabbCc	Aabbcc
aBC	AaBBCC	AaBBCc	AaBbCC	AaBbCc	aaBBCC	aaBBCc	aaBbCC	aaBbCc
aBc	AaBBCc	AaBBcc	AaBbCc	AaBbcc	aaBBCc	aaBBcc	aaBbCc	aaBbcc
abC	AaBbCC	AaBbCc	AabbCC	AabbCc	aaBbCC	aaBbCc	aabbCC	aabbCc
abc	AaBbCc	AaBbcc	AabbCc	Aabbcc	aaBbCc	aaBbcc	aabbCc	aabbcc

Darker skin Same Lighter skin

(a) 20 (b) 27.

2. Environmental influences will alter the colour of a person skin (such as tanning) to different extents.

3. Traits with continuous variation show a normal distribution curve when sampled and a graded variation in phenotype in the population. Such phenotypes are usually determined by a large number of genes and/or environmental influence. Examples include height, weight, hand span, foot size. In contrast, traits with discontinuous variation fall into one of a limited number of phenotypic variants and do not show a normal distribution curve when sampled. Differences in the phenotypes of individuals in a population are marked and do not grade into each other. Such phenotypes are usually controlled by a few different alleles at a few genes, e.g. chin cleft.

4. Student's own plot. Shape of the distribution is dependent on the data collected. The plot should show a statistically normal distribution if sample is representative of the population and large enough.
   (a) Calculations based on the student's own data.
   (b) Continuous distribution, normal distribution, or bell shaped curve are all acceptable answers if the data conform to this pattern.
   (c) Polygenic inheritance: Several (two or more) genes are involved in determining the phenotypic trait. Environment may also have an influence, especially if traits such as weight are chosen.
   (d) A large enough sample size (30+) provides sufficient data to indicate the distribution. The larger the sample size, the more closely one would expect the data plot to approximate the normal curve (assuming the sample was drawn from a population with a normal distribution for that attribute).

270. Changes in a Gene Pool (page 370)

1. This exercise (a)-(c) demonstrates how the allele frequencies change as different of event take place:

Phase 1: Initial gene pool
This is the gene pool before any of the events take place:

	A	a	AA	Aa	aa
No.	27	23	7	13	5
%	54	46	28	52	20

Phase 2: Natural selection

The population is now reduced by 2 to 23. The removal of two homozygous recessive individuals has altered the allele combination frequencies (rounding errors occur).

ISBN: 978-1-927173-94-7

	A	a	AA	Aa	aa
No.	27	19	7	13	3
%	58.7	41.3	30.4	56.5	13.0

Phase 3: Immigration and emigration
The addition of dominant alleles and the loss of recessive alleles makes further changes to the allele frequencies.

	A	a	AA	Aa	aa
No.	29	17	8	13	2
%	63	37	34.8	56.5	8.7

271. Natural Selection (page 371)

1. Stable environments favour the most common phenotypes. Environmental instability provides a chance that extreme phenotypes might be advantageous, but in a stable environment, this is unlikely to happen. Predictability encourages a reduction in phenotypic variability because this is no disadvantage when the environment is largely static.
2. (a) Drought favoured birds with beak sizes at two extremes of the range, since these birds could exploit the small and the large seed sizes.
   (b) Continuation of the drought could lead to further divergence in beak size.
   (c) The selection pressures favouring a bimodal distribution in beak size would be reduced and medium sized beaks would become relatively more common (i.e. there would be a shift from disruptive to stabilizing selection).

272. Selection for Human Birth Weight (page 372)

Sample data and graph below. Note: For the construction of weight classes, it is necessary to have a range of weight categories that do not overlap. The data collected should be sorted into weight classes of: 0.0-0.49, 0.50-0.99, 1.0-1.49, 1.5-1.99, etc.

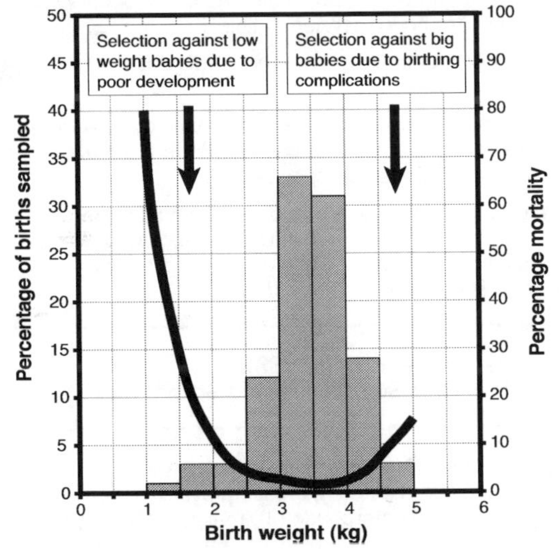

SAMPLE DATA: Use these data if students are unable to collect from local sources

3.740	3.830	3.530	3.095	3.630
1.560	3.910	4.180	3.570	2.660
3.150	3.400	3.380	2.660	3.375
3.840	3.630	3.810	2.640	3.955
2.980	3.350	3.780	3.260	4.510
3.800	4.170	4.400	3.770	3.400
3.825	3.130	3.400	3.260	4.100
3.220	3.135	3.090	3.830	3.970

3.840	4.710	4.050	4.560	3.350
3.380	3.690	1.495	3.260	3.430
3.510	3.230	3.570	3.620	3.260
3.315	3.230	3.790	2.620	3.030
3.350	3.970	3.915	2.040	4.050
3.105	3.790	3.060	2.770	3.400
1.950	3.800	2.390	2.860	4.110
1.970	3.800	4.490	2.640	3.550
4.050	4.220	2.860	4.060	3.740
4.082	3.000	3.230	2.800	4.050
4.300	3.030	3.160	3.300	2.350
3.970	2.980	3.550	3.070	2.715

1. Normal distribution (bell-shaped curve), probably with a skew to the left.
2. 3.5 kg (taken from the table: only 2% mortality)
3. Good correlation. Lowest frequencies of surviving birth weights correspond to birth weights of highest mortality.
4. Selection pressures operate at extremes of the range: Premature babies have reduced survival because their body systems are not fully developed; large babies present problems with delivery as the birth canal can only accommodate babies up to a certain size. Note: Very large babies can occur as the result of gestational diabetes. Before adequate medical intervention, this often led to the death of the mother and/or the baby.
5. Medical intervention can now allow babies that are very premature to survive (babies as small as 1.5 kg have a good chance of survival today, but this has not historically been the case). Caesarean deliveries have also allowed larger babies to be born. Note: This technology is available to wealthy societies thereby reducing the effect of this selection pressure. Developing countries still experience this selection pressure.

273. Selection for Skin Colour in Humans (page 373)

1. (a) Folate is essential for healthy neural development. Note: A deficiency causes (usually fatal) neural tube defects (e.g. spina bifida).
   (b) Vitamin D is required for the absorption of dietary calcium and normal skeletal development. Note: A deficiency causes rickets in children or osteomalacia in adults. Osteomalacia in pregnancy can lead to pelvic fractures and inability to carry a pregnancy to term.
2. (a) Skin cancer normally develops after reproductive age and therefore protection against it provides no reproductive advantage and so no mechanism for selection.
   (b) The new hypothesis for the evolution of skin colour links the skin colour-UV correlation directly to evolutionary fitness (reproductive success). Skin needs to be dark enough to protect folate stores from destruction by UV and so prevent fatal neural defects in the offspring. However it also needs to be light enough to allow enough UV to penetrate the skin on order to manufacture vitamin D for calcium absorption. Without this, the female skeleton cannot successfully support a pregnancy. Because these pressures act on individuals both before and during reproductive age they provide a mechanism for selection. The balance of opposing selective pressures determines eventual skin colouration.
3. Women have a higher requirement for calcium during pregnancy and lactation. Calcium absorption is dependent on vitamin D, making selection pressure on females for lighter skins greater than for males.
4. The Inuit people have such abundant vitamin D in their diet that the selection pressure for lighter skin (for UV absorption and vitamin D synthesis) is reduced and their skin can be darker.
5. (a) Higher chances of getting rickets or (the adult equivalent) osteomalacia due to low UV absorption.
   (b) The simplest option to avoid these problems is for these people to take dietary supplements to increase the amount of vitamin D they obtain.

ISBN: 978-1-927173-94-7

274. Disruptive Selection in Darwin's Finches (page 375)

1. (a) Large and small seeds became relatively more abundant.
   (b) The change in the relative abundance of seed sizes produced a negative selection pressure on finches with intermediate sized beaks. Those with smaller and larger beaks fared better during the drought because they could exploit the smaller and larger seed sizes.
2. Beak size influences fitness, which shows a bimodal distribution. Birds with small beaks (-1.0 single measure) or larger beaks (single measure 1.25) show higher fitness (leave more offspring) than birds with intermediate beak sizes.
3. (a) Mate selection is non-random.
   (b) The graph shows that birds tend to choose mates with a similar beak size. This relationship is stronger when the environmental conditions are more extreme and weaker in more moderate conditions.

275. How Species Form (page 376)

1. (a) Some butterflies rested on top of boulders, others rested in the grass.
   (b) Selection pressure on BSBs is the need to maintain operating body temperature at the high altitude (fitness is higher when they can efficiently absorb heat from boulders). Selection pressure on the GSBs is probably predation as these lowland butterflies survive better where they blend in to avoid detection.

276. Reproductive Isolation (page 377)

1. (a) Geographical isolation is not part of the organism's biology.
   (b) Geographic barriers can be mountains, rivers, lava flows, or oceans. Whether or not it is a barrier depends on the mobility of the organism.
   (c) Geographical isolation enables populations to diverge in response to different selection pressures and (potentially) develop reproductive isolating mechanisms. Reproductive isolation won't usually occur in a population in which there is gene flow (unless by special events such as polyploidy). Note geographic in question should be lower case.
2. Temporal isolation keeps species apart based on when they are physically or reproductively active. A species reproductively active in spring will not be able to mate with a species reproductively active in fall.
3. Courtship displays help members of the same species to clearly identify each other and (through clear signals) assess readiness to mate. It prevents breeding between species as one species will not understand the displays of another.
4. Some orchid flowers are specifically shaped so that access is restricted to a specific pollinator, which pollinates only that species. When the pollinator arrives at the flower of another orchid species it either cannot enter or the pollen may be deposited on the insect in such a way that it won't be transferred successfully to the stigma.
5. Prezygotic isolating mechanisms.

277. Comparing Isolated Populations (page 379)

1.

Mainland	**Nos**	**%**		**Nos**	**%**
Allele A	48	54.5	**Black**	37	84
Allele a	40	45.5	**Pale**	7	16
Total	**88**	100.0			

Island	**Nos**	**%**		**Nos**	**%**
Allele A	12	75	Black	8	100
Allele a	4	25	Pale	0	0
Total	16	100			

2. The frequency of the dominant allele (A) is higher on the island population.
3. Populations that are isolated from each other are often subjected to different selection pressures. This results in gene pool change as different alleles are selected for or against.
4. Block A % for MDH-1 a in order 1-15: 61, 61, 64, 58, 61, 53, 68, 58, 56, 58, 56, 50, 50, 62, 25

   Block B % MDH-1 a in order 1-13: 19, 39, 25, 32, 30, 39, 30, 40, 42, 39, 46, 46, 53
5. Snails, although mobile, are restricted in their need for moisture. The tarmac roads and open areas on the blocks make movement between the two populations difficult as they are open and generally dry areas.
6. The frequencies have become different through a possible founder effect in which the original populations in each block had slightly different allele frequencies to begin with, or random events (i.e. genetic drift) have altered the proportion of alleles being passed to the next generation. Although adjacent, selection pressures in the blocks could also have been subtly different. It is likely that all these processes have contributed to the differences in allele frequencies.
7. Colony 15

278. The Rate of Evolutionary Change (page 381)

1. (a) Punctuated equilibrium: Rapid changes in the physical or biotic (e.g. competitive) environment, e.g. sudden habitat changes or the introduction or removal of other species.
   (b) Gradualism: Gradual, incremental changes in the environment allowing species to track the changes and evolve adaptive features slowly through accumulation of favourable allele combinations.
2. Punctuated equilibrium.
3. (a) Any two of: Sharks, horseshoe crabs, coelacanth.
   (b) Stasis.
   (c) There is no selection pressure for change as they are well adapted (for their niche) as they are (high fitness). There is suggestion that such morphologically static species may track environmental conditions through space and time, and so be subjected to conservative selective pressures throughout their history.

279. Polyploidy and Evolution (page 382)

1. Polyploidy can change the number of sets of chromosomes in an organism, e.g. garlic has 16 chromosomes, while leeks have 32. The change in chromosome number isolates the new species from the parent population as chromosome numbers in gametes are no longer compatible.
2. Bananas, boysenberry, strawberry, garlic, leek, *A. oleraceum,* kiwifruit.
3. Allopolyploidy is the formation of a polyploid by hybridization from different species. Autopolyploidy is the formation of a polyploidy from the same species.
4. Polyploidy in animals tends to disrupt critical systems and produce growth forms that can not be sustained. Plant systems are more tolerant of changes in growth form. Many plant species have been formed from polyploidy.

280. Chapter Review (page 383)

No model answer. Summary is the student's own.

281. KEY TERMS: Did You Get it? (page 384)

1. chiasma (E), chromatid (B), crossing over (F), homologous chromosomes (A), interphase (J), meiosis (C), polyploidy (I), Punnett square (H), recombination (D), speciation (G)
2. (a) A phenotype at the extreme of the phenotypic range has the greatest fitness so the phenotypic norm shifts in the direction of that phenotype, e.g. selection in dark morphs in *Biston* moths in England during the Industrial Revolution.
   (b) Fitness is highest for the most common phenotype and there is selection against phenotypes at the extremes of the phenotypic range, e.g. human birth weights.
   (c) Fitness is lowest for the most common phenotypes and phenotypes at the two extremes of the phenotypic range are favoured, e.g. beak depth in the medium ground finch

ISBN: 978-1-927173-94-7

on Santa Cruz Island during a prolonged drought.

3. Geographic isolation: Populations are isolated by mountains or other geographic barriers. Gene flow is reduced.

   Isolating mechanisms: Isolation of the population leads to differences in mating behaviour and timing of breeding. Continued isolation reduces gene flow even more and results in the development of further prezygotic and postzygotic isolating mechanisms.

   Species isolation: Reduced gene flow between populations results in increased genetic differences over time. Eventually, differences become so great that the species are genetically isolated (true species).

282. Targets for Defence (page 387)

1. (a) The MHC is a cluster of tightly linked genes on chromosome 6 in humans. The genes code for MHC antigens (proteins) that are attached to the surfaces of all body cells and are used by the immune system to distinguish its own from foreign tissue.
   (b) This self-recognition system allows the body to immediately identify foreign tissue e.g. a pathogen, and mount an immune attack against it for the protection of the body's own tissues.

283. The Immune System (page 388)

1. (a) Humoral immune system: Production of antibodies against specific antigens. The antibodies disable circulating antigens.
   (b) Cell-mediated immune system: Involves the production of T cells, which destroy pathogens or their toxins by direct contact or by producing substances that regulate the activity of other cells in the immune system.
2. The presence of an antigen results in the proliferation of specific types of B- and T-cell to target that antigen. Cytokines are also released by macrophages which enhance T-cell activation. More cytokines are released by activated T-cells, and this causes proliferation of other helper T-cells and B-cells.

284. Clonal Selection (page 389)

1. Millions of B-cells form during development. Each B-cell recognises one antigen only, and produces antibodies against it. A pathogen will trigger a response in the B-cell specific that is for it, resulting in proliferation of that B-cell. This is called clonal selection (the antigen selects the B-cell clone that will proliferate).
2. (a) Memory cells: Retain an antigen memory. They can rapidly differentiate into antibody-producing plasma cells if they encounter the same antigen again.
   (b) Plasma cells: Secrete antibodies against antigens (very rapid rate of antibody production).
3. Immunological memory: The result of the differentiation of B cells after the first exposure to an antigen. Those B cells that differentiate into long lived memory cells are present to react quickly and vigorously in the event of a second infection.

285. Antibodies (page 390)

1. Antibodies consist of two heavy (long) peptide chains each attached to a light (short) peptide chain, commonly forming a Y configuration. The variable region at the ends of the heavy and light chains form the antigen binding sites.
2. Antibodies inactivate pathogens in four main ways: Neutralization describes the way in which antibodies bind to viral binding sites and bacterial toxins and stop their activity. Antibodies may also inactivate particulate antigens, such as bacteria, by sticking them together in clumps. Soluble antigens may be bound by antibodies and fall out of solution (precipitation) so that they lose activity. Antibodies also activate complement (a defence system involving serum proteins), tagging foreign cells so that they can be recognized and destroyed.

286. Blood Group Antigens (page 391)

1. See table below:

Blood type	Antigen	Antibody	Can donate blood to	Can receive blood from
A	A	anti-B	A, AB	A, O
B	B	anti-A	B, AB	B, O
AB	A & B	none	AB	A, B, AB, O
O	None	anti-A & anti-B	O, AB, A, B	O

2. If incompatible blood types are mixed, antigens on the RBCs from the donated blood will react with pre-formed antibodies in the blood of the recipient. The donor RBCs will be attacked, clumping together and rupturing.
3. Blood type O^- can be donated to all other blood types (the red blood cells have no antigens). Note that Rh negative blood is relatively rare.
4. Blood type AB^+ can receive blood from any blood type (it has no anti-A or anti-B antibodies). Again, note that Rh negative blood is relatively rare.
5. The discovery of the ABO system allowed blood to be typed before transfusion so that blood could be donated between compatible types (reducing deaths from blood type incompatibility).

287. Allergies and Hypersensitivity (page 392)

1. Histamine mediates the symptoms of hypersensitivity reactions such as inflammation, airway constriction, and itching and watering of the eyes and nose.
2. Being sensitized means to form antibodies to an antigen after exposure to it. Once sensitized, another exposure to the antigen results in an antibody-antigen reaction (and the symptoms of an allergic response).
3. Hypersensitivity is a malfunction of the immune system as it is reacting to particles that are harmless and do not require an immune response.

288. Acquired Immunity (page 393)

1. (a) Immunity as a result of antibodies transferred from one person to another. In this case, the recipient does not make the antibodies themselves.
   (b) Naturally acquired passive immunity arises as a result of antibodies passing from the mother to the fetus/infant via the placenta/breast milk. Artificially acquired passive immunity arises as a result of injection with immune serums e.g. in antivenoms.
2. (a) Newborns need to be supplied with maternal antibodies because they have not yet had exposure to the everyday microbes in their environment and must be born with operational defence mechanisms.
   (b) The antibody "supply" is (ideally) supplemented with antibodies in breast milk because it takes time for the infant's immune system to become fully functional. During this time, the supply of antibodies received during pregnancy will decline.
   (c) Yes. Breast feeding will provide the infant with a naturally acquired passive immunity to help protect it against infections while its immune system develops. Without these acquired antibodies, the infant is more vulnerable to everyday infections against which you already have immunity but he/she does not.
3. (a) Active immunity as a result of the immune response caused by exposure to a microbe or its toxins.
   (b) Naturally acquired active immunity arises as a result of exposure to an antigen such as a pathogen, e.g. natural immunity to chickenpox. Artificially acquired active immunity arises as a result of vaccination, e.g.

ISBN: 978-1-927173-94-7

any childhood disease for which vaccinations are given: diphtheria, measles, mumps, polio etc.

4. (a) The primary response is less pronounced (of smaller magnitude) than the secondary response. The primary response takes longer to develop and is over more quickly than the secondary response, which is rapid and longer lasting.
   (b) The immune system has already been prepared to respond to the antigen by the first exposure to it. When exposed to the same antigen again, it can respond quickly with rapid production of antibodies.
5. (a) Herd immunity refers to the protection that unimmunized people have against a circulating disease by virtue of the fact that most of the population are immunised.
   (b) Once the population contains a high proportion of non-vaccinated people, herd immunity is lost and a circulating disease can spread very rapidly through the community, raising public health costs and contributing to lost productivity.

289. Vaccines and Vaccination (page 395)

1. (a) A vaccine is a preparation of a harmless foreign antigen that is deliberately introduced into the body to produce an immune response.
   (b) The immune response produced by the vaccine is "remembered" by the body and the next time the pathogen is encountered the immune system can respond more quickly to target and destroy the pathogen.
2. Attenuated viruses are more effective in the long term because they tend to replicate in the body, and the original dose therefore increases over time. However, they cannot cause disease because they has been altered to a non-pathogenic form. Attenuated viruses are the closest thing to a natural infection,so elicit strong cellular and antibody responses and often confer lifelong immunity with only one or two doses. Such vaccines are derived from mutations accumulated over time in a laboratory culture, so there is always a risk that they will back-mutate to a virulent form.
3. Generally, bacterial infections can be treated by antibiotics once the infection has taken hold. Viral diseases cannot easily be treated once they have infected a person. Vaccination is therefore important as it prepares the body for a viral infection.
4. (a) If most of the population is immune to the disease, there are too few susceptible hosts to sustain transmission and disease outbreaks stop. The disease can be eradicated because there is no susceptible reservoir in which the pathogen can reside (assuming no non-human reservoir).
   (b) Smallpox eradication was successful because it only affected humans and could not be reintroduced from the wild (unlike viruses such as Ebola). Also, an intense campaign of surveillance and containment was put in place along with high rates of vaccination, especially in highly affected areas.
5. There are persistent pockets of transmission in the four remaining endemic countries. These can transmit the virus to virus free areas. Conflict, political instability, hard-to-reach populations, and poor infrastructure in infected countries increase the difficulty of eradication.

290. Epidemiology (page 397)

1. (a) Introducing the whooping cough vaccine in the 1940s greatly reduced the number of cases of whooping cough contracted each year in the US.
   (b) Whooping cough immunization rates may have dropped due to public perception that the vaccine was unsafe and could cause side effects such as autism.
   (c) Initially, the lower vaccination rate resulted in more cases of whooping cough. The number of whooping cough cases dropped steadily between 2005-2007, before beginning to increase again in 2008.
   (d) Herd immunity will initially provide some protection, but as herd immunity drops (as a result of decreased vaccination rates) the cases of whooping cough will begin to increase.

291. Monoclonal Antibodies (page 398)

1. (a) B-lymphocyte cells.
   (b) The immune system of some people reacts against the foreign (mouse) proteins (antigens).
2. Tumour cells are immortal and can be cultured indefinitely.
3. (a)-(d) in any order:
   (a) Diagnostic tool for detecting pathogens.
   (b) Diagnostic tool for detecting pregnancy hormones.
   (c) Neutralizing endotoxins in blood infections.
   (d) Interfering with T cell activity responsible for transplant rejection.
4. (a) Detection of bacteria or toxins in perishable food would allow the food to be disposed of rather than consumed and hence the possibility of food poisoning avoided.
   (b) Detection of pregnancy at home would give an instant result, and may circumvent a costly visit to a doctor to confirm a pregnancy. For some people, pregnancy detection in the privacy of their home is an attractive option.
   (c) Targeted treatment of cancerous tumours could avoid the need for more invasive or aggressive conventional cancer therapies (which have many side effects).

292. Skeletons and Movement (page 400)

1. Exoskeletons are found on the outside of the organism. They must be shed in order for the organism to grow. Endoskeletons are found inside the body. They are able to grow with the body.
2. The skeleton gives the muscles a rigid attachment point. Jointed skeletons allow movement when muscles contract to move the skeleton about the joint.

293. Movement About Joints (page 401)

1. (a) Joint: a junction between bones that enable skeletal parts to move.
   (b) Ligament: joins bones together at joints providing stability, strength, and flexibility of motion.
   (c) Muscle: contractile tissue that pulls against bones to enable movement of the skeleton.
   (d) Tendon: tough connective tissue structure joining muscle to bone. Resists the tension created by muscle contraction.
2. (a) Pivot: B
   (b) Hinge: D
   (c) Ball and socket: E
   (d) Saddle: A
   (e) Gliding: C
3. The hip joint is a ball and socket joint and allows free movement around a large number of axes with one common centre. The elbow joint is a hinge joint and allows movement in only one plane (flexion/extension).
4. (a) A synovial joint is characterized by the bones forming the joint being enclosed in a synovial cavity, which is filled with lubricating synovial fluid. The articular ends of the bones are covered in cartilage, and ligaments hold the bones in place and give the joint stability.
   (b) Synovial fluid and cartilage jointly act to facilitate free joint movement. Synovial fluid lubricates the joint and reduces friction between the articular cartilage of the bone ends during movement. It also absorbs shocks, and provides nutrients to and removes wastes from the cells of the cartilage. The cartilage cushions the bone ends and, because some synovial fluid is stored within the cartilage, it also provides a store of lubrication.
   Teacher's note: Synovial fluid has a rare property that is becomes more viscous under pressure. It is this property that provides most of its shock absorption qualities.
5. In a cartilaginous joint, bone ends are connected by cartilage and there is limited movement. In a synovial joint, by contrast, the bone ends are separated by a fluid-filled joint cavity and the joint moves freely in one or more planes.

ISBN: 978-1-927173-94-7

294. Antagonistic Muscles (page 403)

1. (a) Prime mover: the muscle primarily responsible for the movement.
   (b) Antagonist: the muscle that opposes the prime mover. i.e. relaxes when prime mover contracts. Its action can be protective in preventing over-stretching of the prime mover during contraction.
   (c) Synergist: assists the prime mover by fine-tuning the direction of limb movement.
2. Muscles can only contract and relax, therefore they can only pull on a bone; they cannot push it. To produce movement, two muscles must act as antagonistic pairs to move a bone to and from different positions.
3. Muscles have an origin on one (less moveable) bone and an insertion on another (more moveable) bone. When the muscle contracts across the joint connecting the two bones, the insertion moves towards the origin, therefore moving the limb. To raise a limb, the flexor (prime mover in this case) contracts pulling the limb bone up (extensor/antagonist relaxed). To lower the limb, the extensor contracts, pulling the limb down (flexor relaxed).
4. (a) Radius (radial tuberosity)
   (b) Ulna
   (c) Triceps (triceps brachii to give it its full name)
   (d) Rotation of ulna and radius
5. (a) The elbow
   (b) The radius and ulna (and hand)
   (c) The brachialis and biceps brachii
6. In an insect leg, the flexor contracts to close the angle between the tibia and upper leg. The extensor contracts to increase the angle between the upper leg and the tibia.

295. Skeletal Muscle Structure and Function (page 405)

1. (a) The banding pattern results from the overlap pattern of the thick and thin filaments (dark = thick and thin filaments overlapping, light = no overlap).
   (b) **I band**: Becomes narrower as more filaments overlap and the area of non-overlap decreases.
   H zone: Disappears as the overlap becomes maximal (no region of only thick filaments).
   Sarcomere: Shortens progressively as the overlap becomes maximal.
2. (a) Relaxed
   (b) The I band can clearly be seen. If the muscle was contracted the I band would not be present.
3. The all-or-none response refers to the way in which an individual muscle fibre contracts maximally (if the stimulus is strong enough) or not at all (if it is not).
4. Changing the frequency of stimulation (more nerve impulses) or changing the number fibres contracting at any one time.

296. The Sliding Filament Theory (page 407)

1. (a) Myosin: Has a moveable head that provides a power stroke when activated.
   (b) Actin: Two protein molecules twisted in a double helix shape that form the thin filament of a myofibril.
   (c) Calcium ions: Bind to the blocking molecules, causing them to move and expose the myosin binding site.
   (d) Troponin-tropomyosin: Bind to actin molecule in a way that prevents myosin head from forming a cross bridge.
   (e) ATP: Supplies energy for flexing of the myosin head (power stroke).
2. (a) ATP (hydrolysis) and Ca^{2+} (to expose the binding sites).
   (b) Ca^{2+} is released from the sarcoplasmic reticulum. ATP is produced by cellular respiration in the mitochondria.
3. Muscle fibres use a lot of ATP during contraction. Having many mitochondria ensures a good ATP supply.

297. Osmoregulation and Excretion (page 408)

1. Osmoregulators regulate fluid balance by regulating or compensating for water fluxes (in and out). Example: Fish (bony and cartilaginous). Osmoconformers do not regulate body fluid composition and allow ion concentrations in their bodies to fluctuate with those of the environment. Example: Marine invertebrates although intertidal organisms may have some limited abilities to osmoregulate and/or tolerate wide salinities.
2. (a) Crabs placed in diluted seawater gained water osmotically because their body fluids are more concentrated than the seawater dilution. The rate of gain was greater in the 50% seawater solution. This is because the osmotic differential is greater than for the 75% dilution.
   (b) This crab species has limited capacity for osmoregulation, i.e. a limited ability to rapidly adjust the composition of its body fluids in a dilute environment.

298. Water Budget in Humans (page 409)

1. Metabolism involves the oxidation of glucose to produce ATP. A by-product of this process is water ($6O_2 + C_6H_{12}O_6 \rightarrow 6CO_2 + 6H_2O$)
2. (a) Dehydration is an excessive loss of water. Overhydration is an excessive intake of water.
   (b) Dehydration results in fatigue and dizziness, mental deterioration, and in severe cases death. Overhydration causes dilution of ions and results in digestive problems, seizures, and possibly coma.

299. Nitrogenous Wastes in Animals (page 410)

1. Main source of nitrogen-containing wastes is the breakdown of nitrogen containing compounds (amino acids and nucleic acids).
2. (a) Uric acid has very low toxicity (it is almost completely non-toxic) and (unlike urea and ammonia) can be excreted with very little water loss. Because very little water is lost in its excretion, uric acid is ideal for terrestrial animals (it reduces their reliance on water availability).
   (b) Ammonia is highly toxic and cannot be held in the body's tissues.
3. The type of excretory product an animal produces is closely correlated with environment. Ammonia is a small, highly soluble molecule. It can be lost rapidly (by diffusion), at low energy cost, through surface contact with the water, where its toxic effects are diluted. It is a common excretory product in aquatic invertebrates (which lose ammonia through the body surface) and the main excretory product in bony fish (which lose it through the gills). Urea is easily soluble and has a moderately low toxicity, so it can be held for some time in solution in the body fluids before excretion. This is a good strategy for terrestrial and semi-aquatic organisms (such as mammals and amphibians) that have (generally) retained a relatively high dependence on fresh water, and have the capacity to carry small volumes of urine along with them. Uric acid prevails in insects, land snails, most reptiles, and birds. All of these are typically terrestrial animals, and the formation of uric acid (which has low toxicity and solubility) can be seen as an adaptation to water conservation in a terrestrial habitat.

300. Excretory Systems (page 411)

1. Uric acid follows active secretion of salts into the Malpighian tubules. Water is absorbed from the waste fluid as it passes through the hindgut. When sufficient water has been withdrawn, the uric acid precipitates out as a paste.
2 (a) Filters the blood and produces the urine.
   (b) Collects the urine from the kidney and conducts it to the bladder.
   (c) Stores the urine.
   (d) Conducts the urine to the outside.

ISBN: 978-1-927173-94-7

301. Kidney Structure (page 412)

1. The kidneys filter the blood of metabolic wastes. By adjusting reabsorption of the filtrate they can maintain the body's internal chemical balance.

2.

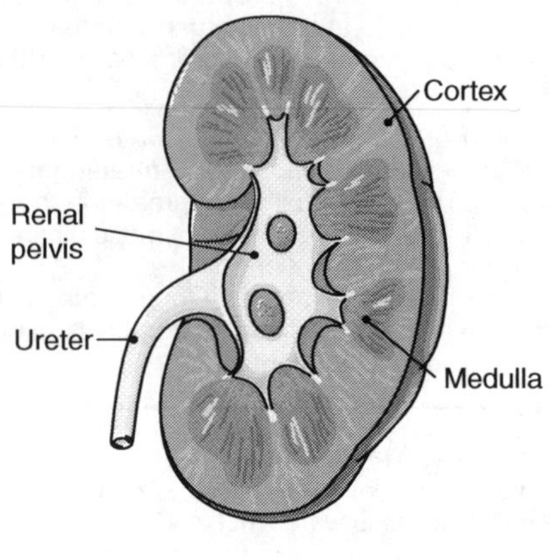

302. Kidney Function (page 413)

1. (a) A nephron is a selective filtering element in the kidney.
   (b) Each nephron is divided into regions where the blood is filtered, and where the filtrate is modified by secretion and reabsorption. By these processes, the nephrons produce the excretory fluid, urine.

2. (a) Glomerular filtration: Produces an initial filtrate of the blood that is similar in composition to blood and can be modified to produce the urine.
   (b) Active secretion: Required answer: Secretion allows the body to get rid of unwanted substances into the urine. Extra detail: Active secretion of chloride in the ascending limb (with sodium following passively) contributes to the maintenance of the salt gradient in the extracellular fluid (this gradient allows water to be reabsorbed in the collecting duct). Secretion of toxins and unwanted ions into the filtrate in the distal tubules allows the blood composition to be adjusted and poisons to be excreted. Energy is used to secrete these substances against their concentration gradients.
   (c) Reabsorption: Essential process that allows the useful substances (required by the body) to be retained from the filtrate (particularly the initial filtrate, where 90% is reabsorbed). The body would waste energy if these substances were not retained.
   (d) Osmosis: Required answer: Osmotic loss of water allows the urine to be concentrated (via loss of water). Extra detail: Osmosis is important in two regions of the nephron: In the descending limb of the loop of Henle, osmotic loss of water concentrates the filtrate so that salt can be withdrawn from the ascending limb to contribute to the salt gradient in the extracellular fluid. In the collecting duct, loss of water by osmosis provides the final concentration of the urine.

3. (a) The salt gradient allows water to be withdrawn from the urine (allows the urine to be concentrated). Note: Because the salt gradient increases through the medullary region, the osmotic gradient is maintained and water can be continually withdrawn from the urine.
   (b) Salt gradient is produced by the active and passive movement of salt from the filtrate into the extracellular fluid in the medulla.

303. The Kidney's Role in Water Conservation (page 415)

1. The longer the loop of Henle, the greater the ability to concentrate urine.

2 (a) For any body mass, medullary thickness is greater in animals in (adapted to) an arid environment.
   (b) The loop of Henle is located in the medulla. The thicker the medulla the longer the loop of Henle. The longer the loop of Henle the more the urine can be concentrated. The more the urine can be concentrated the more water can be conserved and therefore the greater the ability to live in an arid environment.

3 (a)

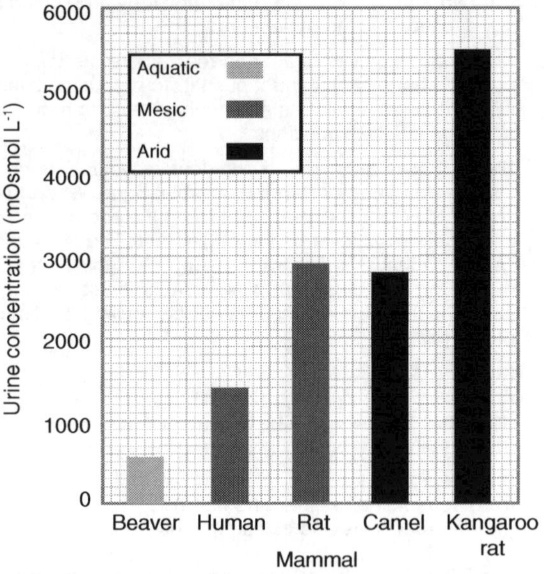

   (b) Kangaroo rat
   (c) Beaver
   (d) Beavers live in or around water and therefore have easy access to it. Kangaroo rats live in deserts with little access to water and have adaptations to conserve water.

4. The metabolism of fat produces water as a waste product and camels store a lot of fat in their humps. They can therefore draw on metabolic water and have greater ability to obtain water than would be expected from their environment alone.

304. Control of Urine Output (page 417)

1. (a) Diabetes insipidus is characterized by excretion of large amounts of very dilute urine (accompanied by a great thirst). The lack of ADH causes excessive urine production (ADH reduces urine output).
   (b) Diabetes insipidus is treated by administering ADH.

2. Alcohol inhibits ADH release causing greater urine output, resulting in dehydration and thirst.

3. (a) Aldosterone increases sodium reabsorption.
   (b) Water follows the salt by osmosis. Blood volume increases as water moves into the blood.

4. The stimulation or inhibition of ADH release results in an adjustment of urine output until homeostasis is restored. The homeostatic adjustment made in response to the ADH release acts back on the hypothalamus to counteract further change.

305. Diagnostic Urinalysis (page 418)

1. Urinalysis is quick and simple to perform and there are diagnostic parameters for particular metabolites. This makes diagnosis and treatment potentially very quick and cost-effective.

2. (a) Diabetes
   (b) The presence of red blood cells in the urine (hematouria) and therefore bleeding in the urinogenital tract.

3. Athletes withhold drug use for a period before testing so that the body has had time to break down and metabolize the drug and excrete it (thereby avoiding detection).

ISBN: 978-1-927173-94-7

306. Kidney Dialysis (page 419)

NOTE: Spelling of dialysate and dialysing solution should follow UK not US spelling as below. It will be corrected in reprints.

1. The dialysing solution is constantly been replaced because it needs to be free of urea and other wastes in order to maintain the concentration gradient between the blood and the dialysate. Without a gradient, urea and other wastes would stay in the blood.
2. Other ions and small molecules do not diffuse into the dialysate because they are at a similar concentration in both blood and dialysate, i.e. the dialysate has a similar ionic composition to blood and there is no concentration gradient for these molecules.
3. The urea passes into the dialysate because there is always less urea present in dialysate than in blood. Therefore there is a concentration gradient for urea between the blood and the dialysate.
4. The general transport process involved in dialysis is diffusion; the movement of molecules from a region of high concentration to a region of low concentration.
5. The dialysing solution runs in the opposite direction to that of the blood (similar to countercurrent heat exchangers and countercurrent flow in fish gills) because in this way almost all of the urea can be removed from the blood in the shortest possible time. Further explanation: The blood at the beginning of the dialysis column (high urea) encounters dialysate that already contains some urea, but there is still a concentration gradient from blood to dialysate. By the time the blood reaches the end of the dialysis column most of the urea has been removed, but it encounters fresh dialysing solution (no urea) and there is still a concentration gradient. Having blood and dialysate running in opposite direction maintains the concentration gradient for a longer time. If the two fluids ran in the same direction the concentrations would quickly equilibrate, and urea would cease to pass into the dialysate.

307. Kidney Transplants (page 420)

1. Acute renal failure (ARF) arises suddenly as a result of infection, blood loss, or dehydration. ARF is characterized by a very sudden drop in urine volume. Chronic renal failure (CRF) develops over months or years as a result of poorly controlled diabetes or high blood pressure, or chronic kidney disease. Unlike ARF, recovery from CRF is not possible.
2. Creatinine is produced at a relatively constant rate and is usually filtered out by the kidneys, so if the levels increase in the blood it would indicate that the kidneys are not operating normally.
3. Kidney transplantation (if the tissue is a good match) offers a high chance of success and a normal life away from the constraints of continual dialysis. Problems include the need to take immunosuppressant drugs and the risk that a suitable donor will not be available when required. Life-time dialysis is expensive, time consuming and cannot help regulate blood pressure in the way that a kidney can. There is also a risk of infection (this is higher with peritoneal dialysis).

308. Animal Sexual Reproduction (page 421)

1. In external fertilization, eggs are released into the environment and fertilized outside of the parent. With this strategy, chance of fertilization is low for any one egg, but increased by synchronized release of very large numbers of eggs and sperm (spawning). This strategy has the added advantage of being of low risk to the breeding individuals. In internal fertilization, eggs are fertilized inside the body of the female by sperm, (usually) deposited in the female reproductive tract by the male. This strategy is advantageous in that the chance of fertilization occurring is high (it involves specifically directed transfer of sperm from male to female via a penis or spermatophore). It has the disadvantage of being a higher risk strategy and often involves energy expense and/or risky courtship and mating behaviours.

5. (a) Internal fertilization/external development: Birds, most reptiles, monotremes (short period of development internally), many sharks.
   (b) Internal fertilization/internal development: Some snakes, some sharks, mammals other than monotremes (marsupial young are still semi- embryonic when born so there is considerable development outside the uterus).
   (c) **Benefit**: Provides a secure, protected environment for the embryo to develop. Allows the gestating parent to continue to supply nutrients until birth.
   Cost: Must maintain the young on the maternal resources, which can compromise the mother if resources are scarce or if she is more vulnerable to predation when pregnant or giving birth.

309. Gametes (page 422)

1. Once deposited in the vagina, sperm must be able to move through the uterus and into the oviducts to reach and fertilize the ovum. Sperm only live for about 48 hours, so they must be able to move quickly to reach the ovum to attempt fertilization before they die.
2. (a) Ova move as a result of the wave-like motion of the cilia in the Fallopian tube.
   (b) Ova must be larger than sperm because they must contain nutritional resources to support the early development of the embryo.
3. The sperm is a motile cell, requiring propulsion to reach the ovum in order to fertilize it. Energy for movement (ATP) is provided by cellular respiration in the large number of mitochondria present in the sperm cell.

310. Spermatogenesis (page 423)

1. (a) Spermatogenesis
   (b) Testis
   (c) Four
   (d) Meiosis
2. LH stimulates synthesis and secretion of testosterone which is required for sperm production. FSH supports the Sertoli cells and sperm maturation.
3. About 100-400 million sperm are needed because only a very small percentage of them eventually reach the ovum (remember that the sperm must negotiate the cervix and then find their way into the Fallopian tube). Note: Although only one sperm fertilizes the egg, the combined action of the enzymes from a large number of sperm is needed in order to help digest the jelly-like barrier around the egg.

311. Oogenesis (page 424)

1. (a) Oogenesis
   (b) In the ovarian follicle within the ovary
2. Sperm are continually produced within the testes from puberty onwards. One spermatogonium produces four identical mature sperm. Many millions of sperm are produced every day. In contrast, a female is born with her full complement of eggs, which develop in stages. The eggs begin to mature from puberty. One egg matures and is released at every menstrual cycle until menopause is reached. Only one mature ovum is produced from an oogonium.
3. Males continue to produce new sperm throughout their reproductive life span. Because the sperm are newly produced, a higher proportion are genetically viable and healthy. Females are born with their full complement of immature eggs. As a female ages so do the eggs, so egg viability declines with age. Oogenesis eventually ceases at menopause.

312. Fertilization and Early Growth (page 425)

1. (a) Capacitation: Changes in the surface of the sperm cell (caused by the acid environment of the vagina) that make possible its adhesion to the oocyte.
   (b) Acrosome reaction: The release of enzymes from the acrosome at the head of the sperm. These enzymes

ISBN: 978-1-927173-94-7

digest a pathway through the follicle cells and the zona pellucida.
(c) Fusion of egg and sperm membranes: Enables the sperm nucleus to enter the egg. The fusion causes a sudden depolarization of the membrane that forms a fast block to further sperm entry.
(d) Cortical reaction: A permanent change in the egg surface that provides a slow (permanent) block to sperm entry. Involves the release of cortical granules into the perivitelline space, followed by the release of substances from the granules that raise and harden the vitelline layer.
(e) Fusion of egg and sperm nuclei: The fusion of nuclei forms the diploid zygote and initiates the rapid cell division that follows fertilization.

2. It is necessary to prevent fertilization of the egg by more than one sperm because this would result in too many chromosomes in the zygote (making the zygote non-viable or unable to survive). Triploidy (a condition resulting from two sperm fertilizing an egg) is frequently found among spontaneous miscarriages.
3. (a) The oocyte is arrested in metaphase of meiosis II after it has undergone the first meiotic division.
(b) Meiotic division proceeds to completion if the egg is fertilized (i.e. fertilization triggers completion).
4. (a) Zygote nucleus: Sperm: 50%, Egg: 50%
(b) Zygote cytoplasm: Sperm: 0%, Egg: 100%
5. Cleavage is the rapid early cell division of the fertilized egg to produce the ball of cells that will become the blastocyst. Cleavage increases the number of cells but not the size of the zygote.
6. (a) Implantation of the blastocyst is important for establishing the close contact between the developing fetus and the uterine lining. The uterine lining can then provide for the early nourishment of the embryo.
(b) HCG prevents degeneration of the corpus luteum, so that it continues to secrete progesterone and maintain the pregnancy (placenta takes over this role at 12 weeks).
7. This the period during which most organ development occurs and the developing tissues are most prone to the damaging effects of drugs.

313. The Placenta (page 427)

1. A double layered, spongy, vascular tissue, formed from both fetal and maternal tissues in the wall of the uterus. The fetal portion of the placenta produces fingerlike projections into the maternal endometrium. These villi contain the capillaries that connect the fetal arteries and vein. The blood vessels of the mother and fetus are in close proximity, forming a complex network that allows for the exchange of nutrients, respiratory gases, wastes, etc.
2. (a) Fetal arteries: Deoxygenated and containing nitrogenous wastes.
(b) Fetal vein: Oxygenated and containing nutrients.
3. The fetal part of the placenta has many villi, rich in capillaries, which project into the endometrium. Nutrients and oxygen diffuse from the maternal arterial blood across capillary walls and into the blood of the fetus. Wastes diffuse in the opposite direction, from the fetal blood into the maternal venules.

314. The Hormones of Pregnancy (page 428)

1. (a) The placenta does not develop fully until ~3 months into the pregnancy. Before this, the progesterone to maintain the endometrium must come from the corpus luteum.
(b) Progesterone and estrogens.
2. (a) Estrogen (high levels) and oxytocin.
(b) Any two of: • Declining progesterone levels. • Loss of placental competency (placenta deteriorates). • High estrogen levels increasing uterine sensitivity to oxytocin. • Peak in oxytocin. • The physiological state of fetus (release of stress hormones etc.) is now known to play a major part in labour onset also.

315. Gestational Development (page 429)

1. There is the general trend of increasing gestational period with increasing mass.
2. Hoofed animals tend to be prey species. They are potentially easy prey when born so must be sufficiently well developed at birth to run with their mother from predators. Herd animals are also often migratory or, at least, constantly on the move. The young must therefore be able to keep up with the herd.
3. (a) Primates have a longer gestation than would be expected from their mass.
(b) Primates tend to have more developed brains in comparison in to body mass than other animals, even at birth. This development takes time.

316. Chapter Review (page 430)

No model answer. Summary is the student's own.

317. KEY TERMS: Did You Get it? (page 432)

1. antibody (E), antigen (A), clonal selection (F), immunity (C), lymphocyte (B), pathogen (G), vaccination (D).

Clarification: definition for antibody should more correctly read " Protein in the blood that identifies and neutralizes foreign material (e.g. bacteria)."
2. (a) Sporadic increases in incidence associated with seasonal changes.
(b) Introduction of vaccination against this pathogen.
3. (a) Joints (c) Antagonistic
(b) Skeletal muscle
4 (a) Kidney (c) Loop of Henle
(b) Nephron (d) ADH, aldosterone
5. Missing words in order: gametes, spermatogenesis, oogenesis, fertilization, placenta, hormones.

318. The Scientific Method (page 434)

1. New data may no longer support the hypothesis or may show it to be flawed or deficient.
New data may explain the system better.
2. Methods that are not repeatable cannot be carried out by others and therefore are not verifiable.
3. In natural environments, such as in field studies, it may be very difficult to control all variables.
4. Prediction 1: That birds will avoid preying on brightly coloured, conspicuous caterpillars. OR

Prediction 2: Naive (inexperienced) birds will learn from a distasteful experience with an unpalatable caterpillar species and will avoid them thereafter. OR

Prediction 3: Birds will prey readily on cryptically coloured caterpillars if these are provided as food. OR
5. (a) Bacterial colonies grow faster at higher temperatures.
(b) Prediction: Bacterial strain A will grow more rapidly at 37°C than at room temperature (19°C).
(c) Set up agar plates of bacteria using sterile techniques. Place 4 plates in a 37°C incubator and 4 on the lab bench (away from any heater unit). Leave all 8 plates for the same length of time (e.g. 24 hours), with all other conditions identical. Evaluate growth by measuring the percentage coverage of the agar plates with bacteria.

319. Variables and Data (page 436)

1. (a) Leaf shape: Qualitative
(b) Number per litter: Quantitative, discontinuous
(c) Fish length: Quantitative, continuous.
2. Quantitative data are more easily analysed in a meaningful way, using descriptive and inferential statistics. Statistical tests are often required to properly test a hypothesis. Note: Qualitative data can be tested statistically but the tests are very complex and generally not very powerful (their ability to

ISBN: 978-1-927173-94-7

detect differences between data sets is sometimes limited).

3. Measure wavelength (in nm) with a spectrophotometer; which measures light intensity as a function of the colour, or more specifically, the wavelength of light.

4. (a) Many variables could be chosen. Examples include: gender, viability (dead or alive), species, presence or absence of a feature, flower colour. These data are categorical; no numerical value can be assigned to them.
   (b) These data are semi-quantitative because arbitrary numerical values have been assigned to a qualitative scale. The numbers are correct in a relative sense, but do not necessarily indicate the true quantitative values.

320. Manipulating Raw Data (page 437)

1. Basic transformations of raw data identify important features (e.g. trends) of the data. Data transformation can identify trends early on and can help a researcher to set the direction for subsequent experimental work.

2. (a) Transformation: Percentage
   Reason: Allows you to compare the relative number of species at each habitat.
   (b) Transformation: Rate
   Reason: Allows you to meaningfully compare data recorded over different periods of time.

321. Planning a Quantitative Investigation (page 438)

1. Aim: To investigate the effect of temperature on the rate of catalase activity.

2. Hypothesis: The rate of catalase activity is dependent on temperature.

3. (a) Independent variable: Temperature.
   (b) 10-60°C in uneven steps: 10°C, 20°C, 30°C, 60°C.
   (c) Unit: °C
   (d) Equipment: A means to maintain the test-tubes at the set temperatures, e.g. water baths. Equilibrate all reactants to the required temperatures in each case, before adding enzyme to the reaction tubes.

4. (a) Dependent variable: Height of oxygen bubbles.
   (b) Unit: mm
   (c) Equipment: Ruler; place vertically alongside the tube and read off the height (directly facing).

5. (a) Each temperature represents a treatment.
   (b) No. of tubes at each temperature = 2
   (c) Sample size: for each treatment = 2
   (d) Times the investigation repeated = 3

6. It would have been desirable to have had an extra tube with no enzyme to determine whether or not any oxygen was produced in the absence of enzyme.

7. Variables that might have been controlled (a-c):
   (a) Catalase from the same batch source and with the same storage history. Likewise for the H_2O_2. Storage and batch history can be determined.
   (b) Equipment of the same type and size (i.e. using test-tubes of the same dimensions, as well as volume). This could be checked before starting.
   (c) Same person doing the measurements of height each time. This should be decided beforehand.

   Note that some variables were controlled: The test-tube volume, and the volume of each reactant. Control of measurement error is probably the most important after these considerations.

8. Controlled variables should be monitored carefully to ensure that the only variable that changes between treatments (apart from the biological response) is the independent (manipulated) variable.

323. Descriptive Statistics (page 441)

1. The modal value and associated ranked entries indicate that the variable being measured (swimmers' height) has a bimodal distribution; the data are not normally distributed. Therefore the mean and median are not accurate indicators of central tendency. Note: the median differs from the mean; also an indication of a skewed (non-normal) distribution.

2. The modal value and associated ranked entries indicate that the variable being measured (sori per frond) has a bimodal distribution i.e. the data are not normally distributed. Therefore the mean and median are not accurate indicators of central tendency. Note also that the median differs from the mean; also an indication of a skewed (non-normal) distribution.

   Erratum for the fern sori example (**first printing only**). The panel to the right of the explanation indicates the **mean** of the data (not the median). We apologize for any confusion caused by this error. It has been corrected in the 2nd printing.

3.

Ladybird mass (mg)	Tally	Total
6.2	I	1
6.7	I	1
7.7	II	2
7.8	I	1
8.0	I	1
8.2	I	1
8.4	I	1
8.8	IIII	4
8.9	I	1
9.8	I	1
10.1	I	1

Median = 8th value when in rank order = 8.4

Mode = 8.8

Mean = 124.7 ÷ 15 = 8.3

Note: To plot a histogram, missing weight classes would have to be included with values of 0.

324. Interpreting Sample Variability (page 443)

1. (a) 496/689 values within ± 1 sd of the mean = 72% (48±7.8, i.e. between 40.2 and 55.8)
   (b) 671//689 values within ± 2 sd of the mean = 97% (48± 15.6, i.e. between 32.4 and 63.6)
   (c) The data are very close to being normally distributed about the mean (normal distribution + 67% of values lie within 1sd of the mean and 95% of values lie between 2 sd of the mean).

2. The mean and the median are very close.

3. N = 30 data set
   (a) Mean = 49.23
   (b) Median = 49.5
   (c) Mode = 38
   (d) Sample variance =129.22
   (e) Standard deviation = 11.37

4. N = 50 data set
   (a) Mean = 61.44
   (b) Median = 63
   (c) Mode = 64
   (d) Sample variance = 14.59
   (e) Standard deviation = 3.82

5. Frequency histogram for the N=50 perch data set.

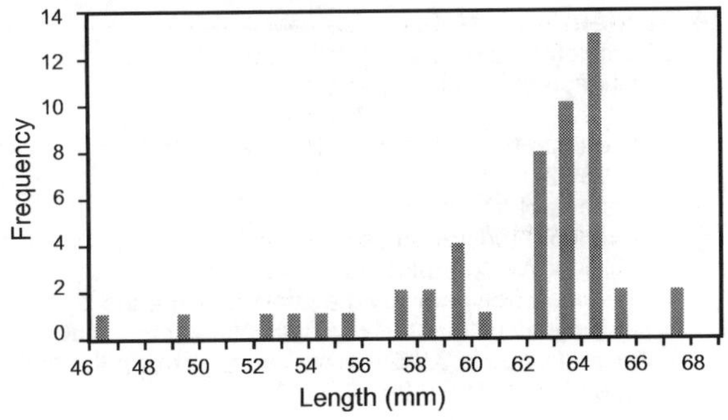

ISBN: 978-1-927173-94-7

Frequency histogram for the N = 30 perch data set.

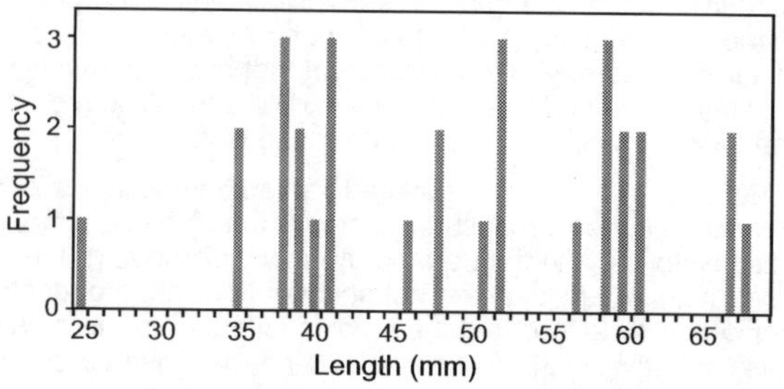

6. (a) The mean and median are very close to each other for the N=30 data set. There is a larger difference between the mean and median values obtained in the N=50 data set.
   (b) The sd obtained for the N=30 set is much larger (11.37) compared to only 3.82 for the larger N=50 data set.
   (c) The N=30 data set more closely resembles the complete data set. The mean and median are quite close to those of the original data set. The mean, median and mode for the N=50 data set are considerably higher than those statistics for the complete data set. The sample variance and standard deviation values for the complete data set fall between those of the two smaller data sets.
7. (a) The frequency histogram for the N=30 data shows a relatively normal distribution. The frequency histogram for the N=50 data shows a non-normal distribution with a skew to the right (negative skew).
   (b) The person who collected the sample in the N=30 set used equipment designed to collect fish randomly. As a result, an unbiased distribution of fish sizes was obtained. Fish collection for the N=50 sample set was biased. The net used did not retain smaller fish, and caught a larger proportion of bigger fish. When plotted on a frequency histogram, the data present as a negative skew.

325. The Student's *t* Test (page 445)

1. (a) The calculated *t* value is less than the critical value of *t* = 2.57. The null hypothesis (of no difference between the control and experimental treatments) cannot be rejected.
   (b) The new *t* value supports the alternative hypothesis at P = 0.05 (reject the null hypothesis and conclude that there is a difference between the control and experimental treatments). The critical value of *t* in this case is 2.23 at 10 d.f. P = 0.05).
2. Outliers can skew a data set, leading to mean values between data sets that are very different (even though the bulk of the data may not be very different). This could result in the rejection of the null hypothesis when it is actually true.
3. Statistical significance refers to the probability that an observed difference (or trend) will occur by chance. It is an arbitrary criterion used as the basis for accepting or rejecting the null hypothesis in an investigation. Teacher's note: In science the term 'significantly different' has a specific meaning. It should not be used in a casual manner when no statistical test has been performed.

326. Student's t test Exercise (page 446)

1. (a) Completed table top of next column
   (b) Variance of population A: 473.57
   Variance of population B: 468.04
   (c) Difference between population means: 135.1
   (d) *t* = 13.92
   (e) Degrees of freedom: 18
   (f) P = 0.05, *t* (critical value) = 2.101
   (g) Decision: We can reject the null hypothesis.
   The difference between the population means is significantly different at P = 0.05. Note at P = 0.001, the critical *t* value is 3.922, so we can also reject the null hypothesis at P = 0.001.

x (counts)		$x - \bar{x}$		$(x - \bar{x})^2$	
Popn A	Popn B	Popn A	Popn B	Popn A	Popn B
465	310	9.3	-10.6	86.5	112.4
475	310	19.3	-10.6	372.5	112.4
415	290	-40.7	-30.6	1656.49	936.36
480	355	24.3	34.4	590.49	1183.36
436	350	-19.7	29.4	388.09	864.36
435	335	-20.7	14.4	428.49	207.36
445	295	-10.7	-25.6	114.49	655.36
460	315	4.3	-5.6	18.49	31.36
471	316	15.3	-4.6	234.09	21.16
475	330	19.3	9.4	372.49	88.36
$n_A = 10$	$n_B = 10$			$\Sigma(x - \bar{x})^2$	$\Sigma(x - \bar{x})^2$
				4262.1	4212.4

The number of samples in each data set

2. (a) Spreadsheet to be completed by the student. Calculated values as follows:
   Mean: A = 455.7, B = 320.6
   Difference between means: 135.1
   Sum of squares: A = 4262.1, B = 4212.4
   Variance: A = 473.567, B − 468.044
   t = 13.92
   (b) New *t* value: 0.76.
   Decision: We cannot reject H0 (P = 0.05). There is no difference between population means.

ISBN: 978-1-927173-94-7